Gene-Mapping Techniques and Applications

edited by

Lawrence B. Schook
Harris A. Lewin
David G. McLaren

University of Illinois at Urbana—Champaign
Urbana, Illinois

CRC Press
Taylor & Francis Group
Boca Raton London New York

CRC Press is an imprint of the
Taylor & Francis Group, an **informa** business

First published 1991 by Marcel Dekker, Inc.

Published 2019 by CRC Press
Taylor & Francis Group
6000 Broken Sound Parkway NW, Suite 300
Boca Raton, FL 33487-2742

First issued in paperback 2019

ISBN-13: 978-0-367-45060-1 (pbk)
ISBN-13: 978-0-8247-8535-2 (hbk)

Visit the Taylor & Francis Web site at
http://www.taylorandfrancis.com

and the CRC Press Web site at
http://www.crcpress.com

Foreword

After several years of controversy, the human genome project is now well under way in the United States and several other countries. It is likely that worldwide efforts in this arena will continue to expand, with the Human Genome Organisation (HUGO) providing a forum for coordination and cooperation. Over the next 15 years several thousand scientists will improve the current human genetic map by almost an order of magnitude. They will make a series of progressively more detailed physcial maps until the 3-billion-base-pair DNA sequence of a human genome is revealed. The wealth of information this project uncovers will change the nature of biomedical research dramatically, as it will provide instant access to any human gene.

It is important to realize that the targets and implications of the human genome project are not limited to *Homo sapiens.* If we already possessed a complete human DNA sequence, our ability to use it today would be very limited. Our ability to analyze raw DNA sequences ab initio to infer aspects of gene function or regulation is very primitive. We need experimental approaches to probe such questions, and most of these require some in vivo studies. For a variety of ethical and practical reasons, the human is a poor subject for such studies. Thus, to interpret the DNA sequence of *Homo sapiens* we need comparative studies on experimental animals. We also need the ability to perform efficient horizontal studies on the human population.

In addition to determining the structure of the human genome, the project must provide the tools to use it. Some of the tools are genome sequences of selected model organisms—bacteria, yeasts, the nematode *Caenorhabditis elegans*, *Drosophila*—and extensive mapping and sequencing work on the mouse. These organisms have been chosen because they represent powerful genetic systems. In addition to assisting the interpretation of the human DNA sequence, they will teach us an enormous amount of basic biology. The most critical tools, however, are massive improvements in the technology needed for genetic mapping, physical mapping, DNA sequencing, and sequence analysis. The experimental tools will have to improve by one to two orders of magnitude if the human genome project is to be completed on schedule and within its projected budget. The computational tools require an improvement of four to six orders of magnitude. Fortunately, in just the first 2 years of the project there have been significant advances on all of these fronts, most notably recognition of the power and versatility of the polyermase chain reaction in numerous genome studies.

The ultimate implications of the genome project go far beyond human biology, and that is why this book is especially timely. The tools of the human genome project, as well as much of the data that will be produced by it, are directly applicable to any organism. It will be especially valuable to apply these tools to domestic animals and economically important crops. In some cases, extensive genetic map data on such organisms already exist. In others, one must start from scratch, and here a number of advances in technology are likely to make the task far easier than it would have been even a year or two ago. These advances include automation of DNA preparation, handling, and analysis; new approaches to polymorphism detection; general ways to find informative polymorphic markers systematically; ways to handle large numbers of samples or markers simultaneously; and more rapid and accurate mathematical algorithms for genetic map construction.

Physical mapping of any organism of interest is already much easier than it was a few years ago. Methods and strategies for manipulating large DNA molecules continue to improve, and new general approaches for constructing coarse maps, or ordering large numbers of clones of DNA fragments, are under development. Automated instruments for DNA sequencing chemistry and sequence data reading are available now; they will surely improve over the next few years, and already in most cases the rate-limiting step is likely to be finding DNA samples worth sequencing rather than the actual sequencing itself.

Thus it is likely that many other genome projects will grow and evolve in parallel with the human genome project. The realities of funding, dictated by the preeminent importance of understanding our own species, dictate that these projects will lag behind the human genome project. It is critical, then, that parallel genome projects be designed flexibly, so that they can adapt to and adopt new technological advances as those occur. One issue that merits considerable thought is how to utilize, most efficiently, genetic and physical map and DNA sequence data on the human and selected model genetic organisms to optimize parallel efforts on other animals and plants. Some major obstacles will need to be overcome. For example, purification of individual human chromosomes is currently much easier than purification of chromosomes from most other species. On the other hand, extensive genetic data already exist on many animals and plants of commercial importance. In some other cases, significant experimental genetic programs could be undertaken now to assist subsequent genome analysis.

Handling and analyzing the massive amounts of data and large numbers of experimental samples that derive from the human genome project is a formidable challenge. General solutions to these problems must be found early in the project; otherwise it will smother in its own products. The methods developed should in no way be restricted to the human species, and thus they will be immediately applicable to all genome efforts. As animal and plant genome studies unfold, there are certain to be fascinating systematic and serendipitous cross-connections with human genome studies. It is impossible to predict now which genes in which organisms will provide great biological insights, but it is certain there will be such insights. For example, simply for direct comparison with the human, swine are undoubtedly closer, and will be more useful in many cases, than mice. Especially where one wants to understand human disease, it is already clear that the phenotype of disease alleles is strikingly modulated by the general physiology of the organism. The disease pathology is not just a product of the biochemistry of a single abnormal gene product. Thus, the genome projects are likely to provide animal analogs or even animal models that will be essential if the breakthroughs of discoveries of genes responsible for many major human diseases are ever to be translated into improved treatment or prevention. In addition, in animals, identification of allelic variants of genes that affect growth, reproduction, lactation, and disease resistance might have a profound impact on the future of international animal agriculture.

Readers of this book will be treated to a broad, thorough coverage of both the methods used in genome studies and their applications. Many should be greatly stimulated to begin to contemplate how a systematic approach to the genome of their particular organism should be initiated.

Charles R. Cantor
Principal Scientist, Department of Energy
Human Genome Project
Lawrence Livermore National Laboratory
Livermore, California

Preface

Identification of the chromosomal coordinates of every human gene is now a realistic goal. The technical advances in nucleic acid biochemistry, enzymology, biophysics, biostatistics, and instrumentation have brought us to the threshold of this new era in the biological sciences. It is anticipated that the benefits of gene-mapping technology and information will extend beyond human medicine to agriculture and other life sciences. The mission of this book is to explain current strategies for mapping genomes of higher organisms and also to explore applications of gene mapping to agriculturally important species of plants and animals. We hope that this book will satisfy those who seek information on state-of-the-art strategies and techniques for gene-mapping research and its applications.

The first section includes four chapters by leading investigators on strategies for gene mapping in plants and animals. Womack gives an overview of mapping strategies currently in use and presents a rationale for the comparative gene-mapping approach. While the basic objective of comparative mapping is to relate chromosomal evolution to speciation, the high level of conservation of linkage relationships across mammalian species allows projection of mapping data from one species to another. The following chapter, by Soller, discusses a theoretical framework and some statistical approaches for mapping genes affecting quantitative traits (QTL), with particular emphasis on livestock populations. The chapter by Hetzel reviews the advantages of establishing a common set of "reference families" for linkage mapping in humans and other species. This particularly useful section explores the power of different family structures for linkage analysis. In the final chapter of this section, Nelson reviews methods for "reverse genetics," i.e., how to go from a genetic marker for a disease gene to localizing it on a chromosome and the eventual cloning and characterization of the genetic lesion. Approaches used for physically isolating the genes responsible for Duchenne muscular dystrophy and cystic fibrosis are contrasted, and new strategies that will employ yeast artificial chromosomes and the polymerase chain reaction are presented.

The chapters in the next section explore in further detail the experimental techniques used for genetic and physical mapping of genes. DNA-sequencing strategies and methods are not considered in this section. In the first chapter, the development of hypervariable genetic markers and their uses in experimental genetics, with particular emphasis given to QTL mapping in domestic animals, is outlined by Georges. Chrisman et al. give a detailed review of in situ hybridization techniques, with a description of the new and powerful methods for high-resolution chromosome banding. Bensaid et al. discuss physical mapping by pulsed-field gel electrophoresis. Included in this chapter is a description of one of the first applications of this technique to a domestic animal species—the physical mapping of genes within the major histocompatibility complex of cattle. The chapter by Fries et al. explores variations of in situ hybridization for monitoring the spacing of markers in species with developing gene maps. Next, Lathrop et al. introduce the concepts and methodologies for statistical evaluation of linkage mapping data, with consideration given to qualitative and quantitative phenotypes.

Current and potential applications of gene mapping are presented next. Beckmann reviews mapping problems and projects with plants, and Watkins discusses identification and use of restriction fragment length polymorphism (RFLP) in the development of human chromosome linkage maps. As a window to the future in animal genetics, Kessler and Schuler present a detailed summary of their studies on the organization and physiological importance of the bovine prolactin gene family. Lewin et al. review the mapping of genes that influence resistance to infectious diseases in humans and domestic animals and discuss experimental designs and strategies for detection and utilization of such genes in livestock species. The final chapter, by Weller and Fernando, describes methodological approaches to achieve the ultimate goal of gene mapping in species of agricultural importance: the improvement of animals through the utilization of genetic marker information. Such methods for "marker-assisted selection" incorporate molecular genetic information into classical animal breeding schemes, thus permitting more accurate selection that is based on genotype as well as phenotype.

We would like to acknowledge the scholarly efforts of our contributors, who have made our task stimulating and rewarding by providing timely and creative articles. Our appreciation also goes to Helen Gawthorp for her assistance in the preparation of this book. The support of our families, colleagues, and students in this effort is gratefully acknowledged.

This book is dedicated to the memory of Dr. C. Larry Chrisman.

Lawrence B. Schook
Harris A. Lewin
David G. McLaren

Contents

Contributors

Jacques S. Beckmann Department of Plant Genetics and Breeding, The Volcani Center, Bet Dagan, Israel

Jonathan E. Beever Department of Animal Sciences, University of Illinois at Urbana—Champaign, Urbana, Illinois

Albert Bensaid Immunology Group, International Laboratory for Research on Animal Diseases, Nairobi, Kenya

Glenn P. Briley Department of Animal Sciences, Purdue University, West Lafayette, Indiana

P. Cartwright Department of Human Genetics, University of Utah, Salt Lake City, Utah

C. Larry Chrisman† Department of Animal Sciences, Purdue University, West Lafayette, Indiana

---

†Deceased

Penelope A. Clamp Department of Animal Sciences, University of Illinois at Urbana—Champaign, Urbana, Illinois

Rohan L. Fernando Department of Animal Sciences, University of Illinois at Urbana—Champaign, Urbana, Illinois

Ruedi Fries Federal Institute of Technology, ETH-Zentrum, Zurich, Switzerland

Michel Georges Genmark Inc., Salt Lake City, Utah

Asoka Gunawardana Federal Institute of Technology, ETH-Zentrum, Zurich, Switzerland

D. J. S. Hetzel Division of Tropical Animal Production, CSIRO, Tropical Cattle Research Centre, Rockhampton, Queensland, Australia

Anita Kaushal Immunology Group, International Laboratory for Research on Animal Diseases, Nairobi, Kenya

Mark A. Kessler Department of Comparative Biosciences, University of Wisconsin—Madison, Madison, Wisconsin

Mark Lathrop Centre d'Etude du Polymorphism Humain, Paris, France

Harris A. Lewin Department of Animal Sciences, University of Illinois at Urbana—Champaign, Urbana, Illinois

Y. Nakamura Cancer Institute, Tokyo, Japan

David L. Nelson Institute for Molecular Genetics, Baylor College of Medicine, Houston, Texas

Lawrence B. Schook Department of Animal Sciences, University of Illinois at Urbana—Champaign, Urbana, Illinois

Linda A. Schuler Department of Comparative Biosciences, University of Wisconsin—Madison, Madison, Wisconsin

Sabina Solinas Federal Institute of Technology, ETH-Zentrum, Zurich, Switzerland

Morris Soller Department of Genetics, The Hebrew University of Jerusalem, Jerusalem, Israel

Alan J. Teale Immunology Group, International Laboratory for Research on Animal Diseases, Nairobi, Kenya

Geoffrey C. Waldbieser Department of Biochemistry, Purdue University, West Lafayette, Indiana

Paul C. Watkins Life Technologies, Inc., Gaithersburg, Maryland

Joel I. Weller The Institute of Animal Sciences, A.R.O., The Volcani Center, Bet Dagan, Israel

James E. Womack Department of Veterinary Pathobiology, College of Veterinary Medicine, Texas A&M University, College Station, Texas

S. Wright Native Plants Inc., Salt Lake City, Utah

John R. Young* Immunology Group, International Laboratory for Research on Animal Diseases, Nairobi, Kenya

---

**Present affiliation*: Agricultural and Food Research Council Institute for Animal Health, Compton, Newbury, Berkshire, England

STRATEGIES FOR GENE MAPPING

1

Strategies and Technologies for Comparative Gene Mapping

James E. Womack

Texas A&M University
College Station, Texas

INTRODUCTION

Comparative gene mapping is the mapping of homologous gene loci in multiple species. The only limitations to comparative mapping, therefore, are the amenability of various species to genomic mapping and the assessment of homology of the genes being mapped. Comparative gene mapping in mammals, the taxonomic class to which this chapter will be restricted, has its roots early in the history of animal genetics. The first genetic linkage reported for a vertebrate species was between albino (*c*) and pink-eye dilution (*p*) in mice (1). Conservation of this linkage relationship was later demonstrated in both rats (2,3) and deer mice (4). In these studies gene homology was based on similarity of phenotypic expression of the respective mutations in three different species. Fortuitously, these mutations occurred and had been preserved in species amenable to laboratory breeding and linkage testing. The interpretation of these data was straightforward: a chromosomal segment of 10–15 map units was conserved in the chromosomal evolution of the three rodent species. Thus, evolutionary conservation of subchromosomal segments in mammals was demonstrated by comparative gene mapping long before banding technologies were sufficiently developed to study genomic evolution on a subchromosomal scale.

Unfortunately, comparative gene mapping lay dormant for a number of years, limited by both the variety of species in which mapping was practical and the paucity of recognizable homologous loci between mammalian species. Mammalian gene mapping was largely limited to mouse mutations that were being kept and studied by a small community of scientists. Asexual mapping protocols emerged in the 1960s and 1970s, however, that permitted gene mapping in previously difficult species, including humans (5), and biochemical genetic techniques (6,7) emerged that clearly identified homologous gene products in different species. Enhanced by the availability of cloned DNA probes in the 1980s, comparative gene mapping has experienced a major revival to become a principal method for the study of mammalian genomic evolution (7–11).

RATIONALE

The primary reason for comparative gene mapping is to identify the boundaries of genomic conservation between evolutionary divergent species. The resolving power of comparative mapping exceeds the resolving power of comparative cytogenetics, even today, in identifying the boundaries of the relatively small segments of conservation between such diverse organisms as mouse and human. When used in a complementary protocol, the two techniques are, however, a powerful tool for the study of chromosome evolution. Combined with comparative mapping high-resolution chromosomal banding has been used to cytologically identify regions of subchromosomal homology between diverse mammalian taxa (12,13).

The ultimate contribution of comparative gene mapping to biological sciences is to provide an understanding of the pathways by which chromosomal evolution has accompanied speciation. A practical contribution, however, is that the knowledge of conserved segments in two different species allows the extrapolation of mapping data from one to the other. For example, the region of mouse chromosome 1 marked by the *Idh*-1, *Fn*-1, and *Cryg*-1 loci contains one or more genes that convey resistance to certain intracellular pathogens in some strains of mice

(14,15). This locus-marked chromosomal segment is conserved in cattle (16,17) and is consequently a "candidate" region for the homologous gene in a species in which polymorphism similar to that in strains of mice could have great economic significance. Comparative mapping can often provide a "first guess" for the regional localization of a gene that has simply been assigned to a chromosome. For example, beta spectrin (SPTB) maps to human (HSA) chromosome 14 (18) and to mouse (MMU) chromosome 12 (19). Its regional assignment on HSA 14 is unknown. A group of other genes that have been regionally assigned to HSA 14q11–q13 map to MMU 14 (Fig. 1). On the other hand, genes on HSA 14q24–q32 map to MMU 12. A first approximation of the regional location of SPTB on the human map is therefore to the proximal half of the q arm with a group of evolutionarily conserved loci. As a note of warning, however, there are sufficient numbers of documented internal rearrangements of conserved sequences between mouse and human to render such regional assignments from comparative data nothing more than tentative preliminary hypotheses.

TECHNOLOGIES

Gene Homology

Comparative gene mapping depends entirely on the accurate assessment of homology between genes of different species. It is sometimes difficult to make this assessment; this issue has been addressed repeatedly in reports of the Committee on Comparative Mapping, an integral part of the biennial Human Gene Mapping (HGM) workshops. The latest committee report from HGM 10 lists the following criteria as standards for the assessment of gene homology (20):

1. Molecular structure
   a. Similar nucleotide or amino acid structure
   b. Similar immunological cross-reaction
   c. Similar subunit structure
   d. Formation of functional heteropolymeric molecules in inter specific cell hybrids in cases of multimeric proteins
   e. Cross-hybridization to the same molecular probe
2. Biological or biochemical function
   a. Similar tissue distribution

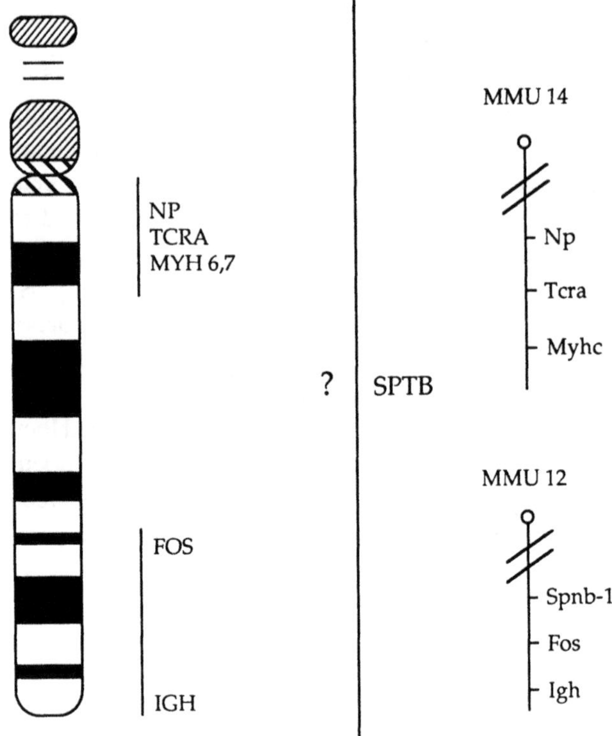

Figure 1 A partial physical map of HSA 14. SPTB is not regionally mapped. Its homolog (Spnb-1) maps to MMU 12, however, suggesting a regional location on the proximal portion of HSA 14q.

b. Similar developmental time of appearance
c. Similar pleiotropic effects
d. Identical subcellular locations
e. Similar substrate specificity
f. Similar response to specific inhibitors

The increased use of heterologous probes has had a major impact on comparative gene mapping. While there is probably not a better single criterion for gene homology than cross-hybridization with a molecular probe, the Committee on Comparative Mapping has recognized an inherent danger in using probes characterized in other laboratories for comparative mapping purposes. Obviously, a mislabeled prove used for mapping can lead to an erroneous comparative chromosomal assignment. The committee strongly recommends that investigators receiving probes from other laboratories perform the tests necessary to verify that the probe being used is the same as the one used in previously published mapping experiments. These criteria include (20):

1. Verification of the restriction sites in the plasmid or phage clone
2. Verification of several restriction fragments after Southern analysis of human or mouse genomic DNA
3. Partial sequence analysis of clones
4. PCR homology with an included oligonucleotide
5. Other traditionally accepted methods of probe verification

If this recommendation is not generally followed, some misassignments are certain to occur and to result in significant problems.

The above discussion assumes the use of biochemical or molecular gene markers for comparative mapping. However, just as albino and pink-eye dilute were identified as homologous mutations in different species (1–4), mutations with similar phenotypic expression are still useful comparative markers. Unfortunately, there does not appear to be an abundance of obvious homologous mutations in different species. The mapping of two similar mutations to a known conserved region in two different species, however, is good preliminary evidence for homology of the mutations.

Mapping Methods

In addition to the assessment of gene homology, comparative gene mapping requires mapping protocols compatible with the species under investigation. Any technique sufficient to localize two genes to the same linkage group, syntenic group, chromosome, or DNA fragment is useful for comparative gene mapping. Consequently, most of the strategies and techniques discussed in this book can be applied to comparative mapping. Some will be more effectively applied to certain species than will others. In situ hybridization, for example, will be relatively ineffective in species whose chromosomes cannot be adequately distinguished. Linkage analysis on the other hand is difficult in species from which large multigeneration families cannot be collected and studied. Nonetheless, advances are rapidly being made in the cytogenetics of a large number of mammalian species, and the reference family approach is making linkage analysis possible for genomes of species that cannot be analyzed for linkage in a single laboratory.

The one technique that is not discussed as a separate chapter in this book, yet has been and probably will continue to be among the most powerful for comparative mapping, is somatic cell genetics. The hybridization of cells from different species (21) and the subsequent selective elimination of chromosomes from one progenitor species (5) is a powerful tool in comparative mapping (Fig. 2). Assignment of genes to syntenic groups is accomplished by the concordant loss or retention of gene products. Electrophoretic differences in homologous enzyme gene products in human and rodent progenitor species was initially used to test for segregation of human genes and the subsequent syntenic mapping of those genes. Cytogenetic analysis of hybrid clones and correlation of the segregation of a particular gene or gene product with a specific human chromosome or chromosome fragment permit the assignment of that gene to a chromosome or chromosomal region. Though still an important end-point of gene identification in somatic cell gene mapping, the electrophoretic separation of gene products is no longer necessary for parasexual genetics. The limitation of mapping only those genes whose products are expressed in cultured cells has been overcome by the use of molecular probes that distinguish host and recipient DNA. Perhaps the most useful mapping tool from the current era of recombinant DNA technology has been the use of Southern blotting for restriction fragment mapping. Already successfully applied to the mapping of several hundred mouse and human genes (20,24,25), this method depends on differences in restriction enzyme sites in the host and donor genomes. As cloned DNA probes to specific genes are developed, this method will have an increasingly greater influence on mammalian gene maps. These methods with numerous refinements have been thoroughly reviewed (22–25). Somatic cell genetics is an ideal method for comparative gene mapping because it circumvents both the requirements of large numbers of offspring from sexual matings and genetic variation within a species. Because of the evolutionary divergence

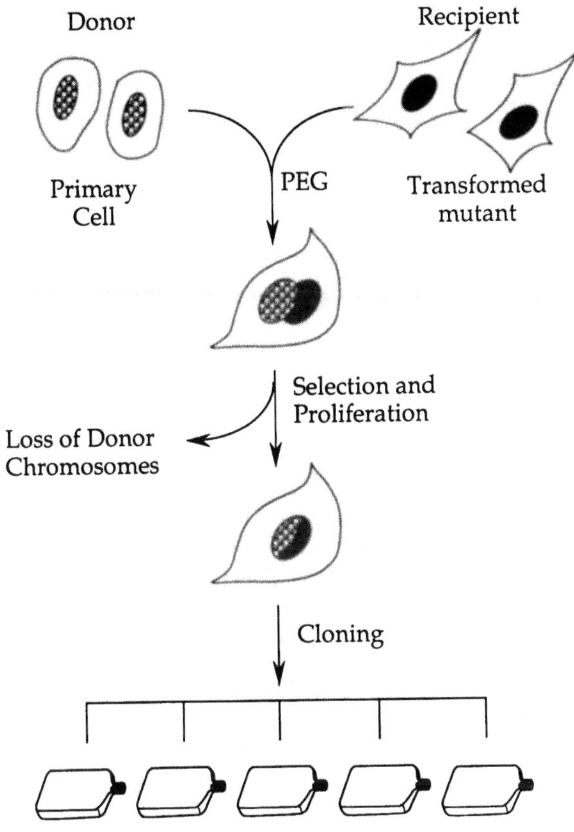

Figure 2 The basic scheme for somatic cell genetic mapping. Synteny of genes is established by concordant loss and retention of genes in the panel of independent hybrid clones.

of such species as humans and mice, the probability of finding electrophoretic differences in their homologous genes and gene products is much greater than the probability of finding variation with a species. While cytogenetic analysis and the assignment of genes to chromosomes is a valuable and powerful aspect of somatic cell genetics, it is not essential for preliminary comparative mapping. Knowing that a group of genes are syntenic in more than one species is an important component of comparative mapping. The relative ease with which hybrid cells can be constructed makes somatic cell genetics an important tool in comparative mapping regardless of the state of cytogenetics of the species being mapped.

Cell hybridization facilitated by Sendai virus or polyethylene glycol has been the principal method for transferring donor chromosomes into recipient cells. Other methods have been developed, however, for specific experimental strategies (25). The isolation of donor chromosomes into microcells (26) provides experimental control over chromosome segregation and allows the transfer of one or a few intact donor chromosomes which are degraded into chromosomal fragments after endocytosis (27). Eukaryotic cells can also be transformed by purified DNA (28), either by endocytotic uptake of calcium-precipitated DNA or by direct injection into the nucleus with micropipettes. The size of the donor DNA molecule can be specifically controlled in these experiments and is often 50 kilobases or smaller. Each of these methods provides a means of transferring progressively smaller segments of donor chromosome into recipient cells and consequently progressively extending the limits of refinement of gene mapping by parasexual methods. Physical ordering of genes on chromosomes is now being effected by the use of radiation hybrids, another modification of the classical somatic cell technique (29). In the radiation-fusion technique, the chromosomes of the donor cell line are fragmented by irradiation prior to fusion. Genes can be ordered by the relative frequencies at which markers are cotransferred from the irradiated donor line to the established rodent recipient line. If the donor is itself a chromosomally reduced hybrid cell line (one or two donor chromosomes present), the protocol can be particularly definitive (30).

STRATEGIES

Gene maps of 29 different mammalian taxa were summarized and compared at the latest Human Gene Mapping Workshop (20). The status of the maps of 23 eutherian species is summarized in Table 1. These numbers are, of course, rapidly being outdated. The cow map, for example, has grown to 160 loci on all 29 autosomal syntenic groups since this

Table 1 Status of Gene Maps of Eutherian Mammals as of HGM10 (June 1989)

Species	Haploid number	Known linkage and syntenic groups	Number of mapped genes
Human	23	24	1600+
Chimpanzee	24	21	70
Gorilla	24	23	44
Orangutan	24	17	38
Gibbon	26	21	43
Rhesus	21	15	39
Baboon	21	16	36
African green	30	17	35
Capuchin	27	14	26
Aotes	25	15	59
Marmoset	23	16	27
Mouse lemur	33	14	35
Mouse	20	21	1450+
Rat	21	16	128
Chinese hamster	11	11	63
Cat	19	18	64
Rabbit	22	17	55
Dog	39	26	47
Cow	30	26	116
Sheep	27	16	35
Pig	19	17	40
Horse	32	5	21
American mink	15	14	41

Source: Taken from the Report of the Committee on Comparative Mapping (20).

report was published. It is obvious, however, that the maps of *Homo sapiens* and *Mus musculus* far exceed all others and, for that reason, should be the prototypic maps for comparative mapping in other species. The human genome initiative has targeted the mapping and complete sequencing of the human genome over the next 15 years. Numerous genes whose homologues are of interest in other species will consequently be identified and mapped first in *Homo sapiens*. The mouse map also continues to expand, thanks to the availability of inbred strains, congenic strains, recombinant-inbred strains, and, more recently, offspring from interspecific hybrids. This battery of genetically defined stocks and strains will continue to help define the genetic basis of complex traits and, in the process, facilitate the mapping of the genes involved. It behooves those of us with interests in the genomes of other species to pay careful attention to the rapidly developing prototypic maps of mouse and human. Moreover, the maps of mouse and human can and will continue to benefit the development of each other. Part of the strategic plan to develop a map of any mammalian genome should be the mapping of loci whose homologs span the human and mouse genomes at regular intervals. Evolutionary breakpoints will be identified as boundaries of conserved linkage or synteny.

The rapidly developing bovine gene map has drawn significantly from the mouse and human maps. As evident from Table 2, 142 homologous genes have been mapped in both cattle and humans and most are mapped in all three species. The boundaries of evolutionary conservation of several chromosomal segments are emerging from this study. HSA 21, the Down syndrome chromosome, for example, appears to be conserved intact in cattle while its genes are dispersed over at least three mouse chromosomes. HSA 9 and 12 are also almost completely conserved in cattle loosing only the q terminal region in each instance to a different bovine syntenic group. The same human genes are spread over three and three mouse chromosomes, respectively.

SUMMARY

Comparative gene mapping is a rapidly developing discipline. It is not only a tool to understand chromosomal evolution, but it provides shortcuts to mapping new genes in species for which chromosomal conserva-

Table 2 Table of Homologous Loci Mapped in Humans, Cattle, and Mice, Arranged by Human Chromosomal Location

	Chromosome location		
Gene	Human	Cow	Mouse
GDH	1p	5	3
PGD	1p	U1	4
ENO1	1p	U1	4
PGM1	1p	U6	4
AMY	1p	U6	3
AT3	1q	U1	1
ABLL	1q	U1	1
REN	1q	U1	1
IDH1	2q	8	1
CRYG	2q	8	1
FN1	2q	8	1
VIL1	2q	8	1
ACY1	3p	U12	9
GPX	3	U12	
RHO	3q	U12	6
TF	3q	U10	9
CP	3q	U10	9
SST	3q	U10	16
GAP43	3q	U10	16
PGM2	4	6	5
PEPS	4	6	5
IGJ	4q	6	5
ADH2	4q	6	3
IF	4q	6	
IL2	4q	U23	3
FGB	4q	U13	3
FGG	4q	U13	3
SPARC	5q	U22	11

Table 2 Continued

Gene	Chromosome location		
	Human	Cow	Mouse
PRL	6p	23	17
CYP21	6p	23	17
HSPA1	6p	23	17
GLO	6p	23	17
HLA	6p	23	17
MUT	6p	23	17
ME1	6q	U2	9
PGM3	6q	U2	9
SOD2	6q	U2	17
TCP1	6q	23	17
HOX1	7p	U13	6
IL6	7p	U13	5
MDH2	7	U8	5
ASL	7	U8	5
GUSB	7q	U8	5
COLIA2	7q	U13	6
MET	7q	U13	6
NEFL	8p	U18	
GSR	8p	U14	8
PLAT	8p	U14	8
MOS	8q	14	4
CYP11B	8q	14	
CA2	8q	14	3
TG	8q	14	15
MYC	8q	14	15
IFNA	9p	U18	4
IFNB	9p	U18	4
GGTB2	9p	U18	4
ACO1	9	U18	4
ALDOB	9q	U18	

Table 2 Continued

Gene	Chromosome location		
	Human	Cow	Mouse
ALDH1	9q	U18	19
GSN	9q	U18	
ITIL	9q	U18	
CS	9q	U18	2
GRP78	9q	U16	
AK1	9q	U16	2
ABL	9q	U16	2
ASS	9q	U16	2
IL2R	10p	U11	
VIM	10p	U11	2
ADRA2R	10q	U26	
GOT1	10q	U26	19
CYP17	10q	U26	
PTH	11p	15	7
HBB	11p	15	7
LDHA	11p	U7	7
CAT	11p	15	2
FSHB	11p	15	2
CRYA2	11	15	
TYR	11q	U7	7
GAPD	12p	5	6
TPI1	12p	5	6
A2M	12p	5	6
LDHB	12p	5	6
KRAS2	12p	5	6
HOX3	12q	5	15
GL1	12q	5	15
INT1	12q	5	15
LALBA	12q	5	15
KRTB	12q	5	15

Table 2 Continued

Gene	Human	Cow	Mouse
	Chromosome location		
NKNB	12q	5	
LYS	12q	5	
PEPB	12q	5	10
PAH	12q	5	10
PAH	12q	5	10
IGFI	12q	5	
IFNG	12q	5	10
ALDH2	12q	U23	4
ESD	13q	U13	14
NP	14q	10	14
FOS	14q	10	12
IGH	14q	21	12
BM2	15q	10	2
PKM2	15q	10	9
HEXA	15q	10	9
MPI	15q	21	9
CYP11A	15q	21	9
FES	15q	21	7
HBA	16p	U8	11
PRM	16	U8	16
DIA4	16q	18	
POLR2	17p	U27	11
HOX2	17q	19	11
GH	17q	19	11
KRTA	17q	19	11
YESI	18q	U28	
MYB	18q	19	18
PEPA	18q	19	18

Table 2 Continued

Gene	Chromosome location		
	Human	Cow	Mouse
AMH	19	U22	
GPI	19q	18	7
ITPA	20p	U11	2
HCK	20q	U11	2
SRC	20q	U11	2
ADA	20q	U11	2
IFREC	21q	U10	16
APP	21q	U10	16
SOD1	21q	U10	16
PRGS	21q	U10	16
ETS2	21q	U10	16
PAIS	21q	U10	16
COL6A1	21q	U10	10
COL6A2	21q	U10	10
CBS	21q	U10	17
CRYA1	21q	U10	17
PFKL	21q	U10	
S100B	21q	U10	10
CD18	21q	U10	10
IGL	22q	U23	16
DMX	X	X	X
PGK1	X	X	X
GLA	X	X	X
HPRT	X	X	X
F9	X	X	X
G6PD	X	X	X

tion with a prototypic map has been demonstrated. Much of the emerging data from the human and mouse maps can be extrapolated to the underdeveloped maps of other species if the boundaries of conservation of chromosomal segments can be established. Biochemical and molecular gene markers are excellent tools for establishing gene homology if the appropriate criteria are met. While a variety of mapping methods are applicable to different species, the hybrid somatic cell genetic method remains unsurpassed for efficiently providing an overall estimate of the extent of genomic conservation between species. Comparative mapping to date indicates that mammalian genomes in general are highly conserved and that the developing new maps of mammalian species can draw from one another and especially from the prototypic maps of mouse and human.

REFERENCES

1. Haldane JBS, Sprunt AD, Haldane NM. Reduplication in mice. J Genet 1915; 5:133–135.
2. Castle WE, Wachter WL, Variation of linkage in rats and mice. Genetics 1924; 9:1–2.
3. Feldman HW. Linkage of albino allelomorphs in rats and mice. Genetics 1924; 9:487–492.
4. Clark FH. Linkage of pink-eye and albinism in the deer mouse. J Hered 1936; 27:256–260.
5. Ruddle FH. Linkage analysis using somatic cell hybrids. Adv Hum Genet 1972; 3:173–235.
6. Markert CL. The molecular basis for isozymes. Ann NY Acad Sci 1968; 151:14–40.
7. Womack JE. Linkage of mammalian isozyme loci: a comparative approach. In: Rattazzi MC, Scandalios JG, Whitt GS, eds. Isoenzymes. current topics in biological and medical research, Vol 6. New York: Alan R. Liss, 1982:207–245.
8. Searle AG. Comparative and historical aspects of the mouse genome. In: Berry RJ, ed. Biology of the house mouse. London: Academic Press, 1981:63–84.
9. Nadeau JH, Tayler B. A. Lengths of chromosomal segments conserved since divergence of man and mouse. Proc Natl Acad Sci USA 1984; 81:814–818.

10. Womack JE. Comparative gene mapping: a valuable new tool for mammalian development studies. Dev Genet 1987; 8:281-293.
11. O'Brien SJ, Seuanez NH, Womack JE. Mammalian genome organization: an evolutionary view. Annu Rev Genet 1988; 22:323- 351.
12. Nash WG, O'Brien SJ. Conserved subregions of homologous G-banded chromosomes between orders in mammalian evolution: carnivores and primates. Proc Natl Acad Sci USA 1982; 79:6631-6635.
13. Sawyer JR, Hozier JC. High resolution of mouse chromosomes: banding conservation between man and mouse. Science 1986; 232:1632- 1635.
14. Bradley DJ. Regulation of *Leishmania* populations within the host. II. Genetic control of acute susceptibility of mice *Leishmania donovani* infection. Clin Exp Immunol 1977; 30-130-140.
15. Skamene E, Gros P, Forget A, Patel PJ, Nesbitt MM. Regulation of resistance of leprosy by chromosome 1 in the mouse. Immunogenetics 1984; 19:117-124.
16. Adkison LR, Skow LC, Thomas TL, Petrash M, Womack JE. Somatic cell mapping and restriction fragment analysis of bovine genes for fibronectin and gamma crystallin. Cytogenet Cell Genet 1988; 47:155-159.
17. Zneimer SM, Womack JE. Regional localization of the fibronectin and γ-crystallin genes to bovine chromosome 8. Genomics 1989; 5-215-220.
18. Watkins PC, Eddy R, Winkelmann JC, Forget BG, Shows TB. Assignment of the gene for beta-spectrin (SPTB) to human chromosome 14. Cytogenet Cell Genet 1987; 46:712.
19. Laurila P, Cioe L. Kozak CA, Curtis PJ. Assignment of mouse beta spectrin gene to chromosome 12. Somat Cell Mol Genet 1987; 13:93-98.
20. Lalley PA, Davisson MT, Graves JAM, O'Brien SJ, Womack JE, Roderick TH, Creau-Goldberg N, Hillyard AL, Doolittle DP, Rogers JA. Report of the committee on comparative mapping (HGM10). Cytogenet Cell Genet 1989; 51:503-532.
21. Weiss MC, Ephrussi B. Studies of interspecific (rat x mouse) somatic hybrids. II. Lactic dehydrogenase and β-glucuronidase. Genetics 1966; 54:1111-1122.
22. Ruddle FH, Creagan RP. Parasexual approaches to the genetics of man. Annu Rev Genet 1975; 9:407-486.
23. Shows TB Genetic and structural dissection of human enzymes and enzyme defects using somatic cell hybrids. In: Rattazi MC, Scandalio JG, Whitt GS, eds, Isozymes. Current topics in biological and medical research, Vol 2. New York: Alan R. Liss, 1977: 107-158.

24. Shows TB, Sakaguchi AY, Naylor SL. Mapping the human genome, cloned genes, DNA polymorphisms, and inherited disease. Adv Hum Genet 1982; 12:341–452.
25. Ruddle FH. A new era in mammalian gene mapping: somatic cell genetics and recombinant DNA methodologies. Nature 1981; 294:115–120.
26. Fournier REK, Ruddle FH. Microcell-mediated transfer of murine chromosomes into mouse, Chinese hamster, and human somatic cells. Proc Natl Acad Aci USA 1977; 74:319–323.
27. Klobutcher LA, Ruddle FH. Chromosome mediated gene transfer. Annu Rev Biochem 1981; 50:533–554.
28. Wigler M, Silverstein S, Lee L-S, Pellicer A, Cheng Y-C, Axel R. Transfer of purified herpes virus thymidine kinase to cultured mouse cells. Cell 1977; 11:223–232.
29. Goss SJ, Harris H. Gene transfer by means of cell fusion. I. Statistical mapping of the human X-chromosome by analysis of radiation-induced gene segregation. J Cell Sci 1977; 25:17–37.
30. Cox DR, Pritchard CA, Uglum E, Casher D, Kobori J, Myers RM. Segregation of the Huntington disease region of human chromosome 4 in a somatic cell hybrid. Genomics 1989; 4:397–407.

2

Mapping Quantitative Trait Loci Affecting Traits of Economic Importance in Animal Populations Using Molecular Markers

Morris Soller

The Hebrew University of Jerusalem
Jerusalem, Israel

Determination of linkage between marker loci and quantitative trait loci (QTL) will depend on the presence of linkage disequilibrium between alleles at the marker locus and alleles at the QTL. This disequilibrium will generate marker-associated quantitative effects that can be detected by appropriate statistical analyses.

Linkage disequilibrium can be produced by crossing populations that differ in allele frequencies at marker loci and/or QTL. Within a segregating population at linkage equilibrium, linkage disequilibrium will inevitably be found within a given family as a result of the specific coupling relationships between marker alleles and QTL alleles in the parents of the family. Linkage disequilibrium can also be present in a segregating population as a result of genetic drift, founder effects, or selection.

Following detection of linkage between marker and QTL, map distance between marker and QTL and more exact location of QTL on a genetic map can be determined by application of maximum-likelihood methods or by use of pairs of markers bracketing a QTL between them.

Marker-associated quantitative effects are generally detected by least-squares linear-models methods. In some cases, however, maximum-likelihood approaches can increase the power of marker-QTL linkage determination. Similarly, when a QTL is bracketed between a pair of flanking marker loci, additional power and information can be obtained by considering the quantitative value of parental and recombinant marker genotypes separately. When information on quantitative traits is routinely collected for large numbers of individuals, information on marker-QTL linkage for a given sample can sometimes be markedly increased by carrying out marker analyses on selected individuals only. Along the same lines, information on quantitative trait value of progeny can be utilized to increase power of market-QTL linkage analyses carried out on the parent generation.

INTRODUCTION

The first experiment designed to map loci affecting QTL through linkage to mendelian marker loci was carried out by Sax (1923). Application of such procedures to agricultural populations was limited for many years by lack of suitable markers. With the advent of immunological, biochemical, and more recently, DNA-level markers, this limitation has for the most part been removed. Large numbers of polyallelic markers, either of the VNTR type, uncovered by RFLP techniques (Nakamura et al., 1987), or of the microsatellite type, uncovered by use of the polymerase chain reaction (Weber and May, 1989; Litt and Luty, 1989), are now available. Along with this has come the development of efficient experimental designs and modes of analysis for the determination of linkage between marker loci and QTL. Together these make it possible to aim at mapping most of the loci affecting traits of economic importance in animal populations. This map information forms the basis for marker-assisted selection programs (MAS). Theoretical analyses of MAS show that such programs may be able to add significant increments to the rate of genetic progress for a variety of traits and organisms and may make it possible to attempt breeding programs that would otherwise be impractical (see Chapter 14, this volume). Mapping of QTL will also be a preliminary step to their cloning by way of reverse genetics (Rommens et al., 1989; Kerem et al., 1989; Suthers et al., 1989) and may facilitate the physiological and developmental analysis of traits of economic importance by identifying and isolating their individual biological components. In this chapter experimental design and analyses for determining linkage between marker loci and QTL and for determining map location of QTL and their genetic effects in animal populations are presented (see also Chapter 10, this volume).

SOME GENERAL CONSIDERATIONS AND ASSUMPTIONS

In all cases, determination of linkage between marker loci and QTL will depend on the presence of linkage disequilibrium between alleles at the marker locus and alleles at the QTL. This disequilibrium will generate marker-associated quantitative effects that can be detected by appropriate statistical analyses. The expected degree of linkage disequilibrium between marker and QTL will depend in a nonlinear manner on the distance between marker and QTL (as well as on the prior history of the population). Consequently, power of a marker-QTL linkage experiment will be a sensitive function of the number of markers scored per individual, since this will determine the average distance between marker and QTL. Clearly, power will also depend on the number of individuals evaluated in a given experiment. The product of number of markers scored per individual and number of individuals evaluated per experiment will therefore determine the total number of marker determinations in a given marker-QTL linkage experiment. Because of the nonlinear relationship of disequilibrium to marker-QTL distance, there will be optimum spacing of markers for a marker-QTL linkage experiment (in the sense of least number of marker determinations for given overall power). There will also be a marker spacing that minimizes costs of the experiment. This will depend, in addition to the above, on the relative cost of raising and evaluating an animal with respect to quantitative traits and of obtaining a DNA sample, as compared to the cost of carrying out a marker determination (Darvasi and Soller, 1990).

In all that follows, it will be assumed that a market-associated quantitative effect is due to a single QTL in linkage to the marker. In actual fact, in a greater or lesser proportion of the cases, a marker-associated quantitative effect may be due to segregation at more than one QTL in the vicinity of the marker. Using current mapping methods, it is not possible to distinguish between a single QTL or a cluster of QTLs within a limited chromosomal region. For many breeding purposes clustering will not materially affect the usefulness of information on marker-associated quantitative effects, since the object of QTL mapping will be to identify chromosomal regions that have economically significant effects on traits of agricultural importance, so that these chromosomal regions can be manipulated in breeding programs by means of the associated markers. Clustering of QTL will, however, have the effect of biasing map distances (McMillan and Robertson, 1974) and hence will render problematic reverse genetic approaches based on supposed map locations of identified QTL. The extent to which segregating QTL affecting a given quantitative trait are clustered is presently unknown and is a major scientific question that remains to be clarified.

Marker-QTL linkage analyses will often involve a large matrix consisting of many markers tested against a number of traits. Since markers will exhibit linkage with one another and traits may be correlated to a greater or lesser extent, determining appropriate levels of statistical significance for the entire matrix of results is a problem. This has been considered in some detail for the case of a single trait mapped with respect to many markers (Lander and Botstein, 1989).

THEORY AND SOME NUMERICAL RESULTS

Linkage Disequilibrium

Populations of agricultural animals invariably reproduce by cross-fertilization. For such populations, the particular source of linkage disequilibrium utilized for marker-QTL mapping will depend on the genetic history of the trait and populations involved. Four situations can be distinguished:

1. Crosses between inbred lines that are at fixation for alternative alleles at both marker loci and QTL.
2. Individual families within a single population segregating for both marker loci and QTL.
3. Crosses between populations that are segregating with respect to markers, but are at or close to fixation for alternative alleles at QTL affecting a trait of interest.

4. Overall linkage disequilibrium within a population segregating for both marker loci and QTL. This would be particularly appropriate for populations that have been maintained with small effective numbers for many generations e.g., some of the smaller cattle or poultry breeds.

Each of these situations will now be considered in detail with respect to the degree of linkage disequilibrium that can be anticipated, the expected magnitude of the marker-associated quantitative effect due to linkage disequilibrium, the type of analysis that will be applied to uncover the marker-associated effect, and the power of the analysis in the sense of the experimental size required to uncover typical marker-associated effects.

Crosses Between Inbred Lines

When inbred lines are available that differ with respect to marker loci and also with respect to QTL affecting a trait of interest, marker-associated effects can be readily detected by crossing the lines and comparing quantitative value of alternative marker genotypes in the F_2 or BC generations (Soller et al., 1976). This class of experiment is of primary interest in plant genetics, where numerous species are selfers and inbred lines are plentiful. In animals; this would include experimental studies involving the available inbred lines of mice or chickens.

Using the symbols **M/m** to represent alleles at the marker locus and **A/a** to represent alleles at the QTL, a biometrical analysis of the F_2 of a cross between two inbred lines (**MA/MA x ma/ma**) is shown in Figure 1. Codominance, typical for isozyme or DNA-level markers, is assumed at the marker locus, so that all three marker genotypes can be distinguished. As a result of linkage, F_2 individuals having the **MM** marker genotype will include primarily **AA** genotypes at the QTL, while F_2 individuals having the **mm** marker genotype will include primarily **aa** genotypes at the QTL. This will generate a marker associated quantitative effect that can be detected by least-squares methods, e.g., by a *t*-test of analysis of variance (ANOVA).

The analysis also shows, however, that as a result of recombination, the homozygous marker genotypes in the F_2 generation each represent a mixture of genotypes at the QTL. Consequently, letting 2d equal the difference in mean quantitative value between alternative homozygous

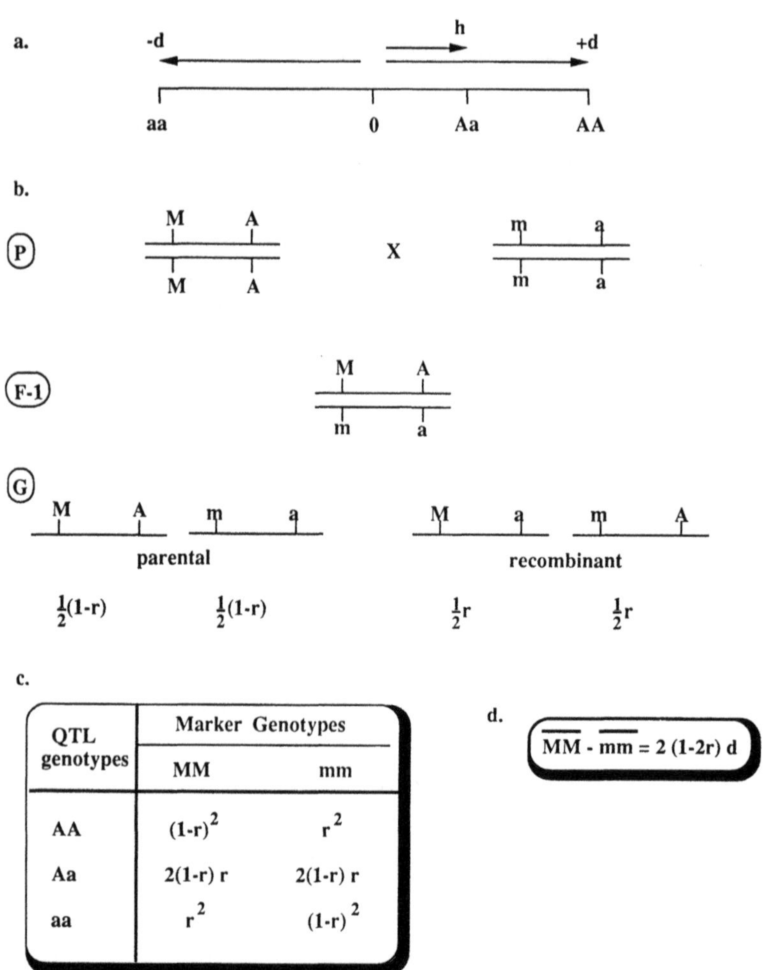

QTL genotypes	Marker Genotypes MM	mm
AA	$(1-r)^2$	r^2
Aa	$2(1-r)r$	$2(1-r)r$
aa	r^2	$(1-r)^2$

Figure 1 Linkage between marker locus and QTL in the F_2 generation of cross between inbred lines. **M** and **m**, alleles at the marker locus; **A** and **a**, alleles at the QTL; d and h, main effect and dominance effect at QTL; r, proportion of recombination between marker locus and QTL; $\overline{\mathbf{MM}}$ and $\overline{\mathbf{mm}}$, mean value of marker genotypes in the F_2 generation. (a) Mathematical scale showing gene effects at the QTL. (b) Diagrammatic chromosomes of parents, F_1 progeny, and gametes formed by F_1, showing coupling between marker alleles and QTL alleles and frequency of parental and recombinant F_1 gametes. (c) Relative genotype frequencies at the QTL, within F_2 progeny having alternative homozygous marker genotypes. (d) Difference between mean values of alternative homozygous marker genotypes in the F_2 generation.

genotypes at the QTL (i.e., **AA** – **aa** = 2d), the difference in mean quantitative value between alternative homozygous linked marker genotypes will be attenuated by recombination between marker and QTL. Thus, main effect at the QTL (d) and proportion of recombination (r) are confounded in the observed difference between quantitative value of marker genotypes. For a backcross population, the comparison would be between **Mm** and **mm** marker genotypes, and the expected difference in quantitative value between marker genotypes would be **Mm** – **mm** = (1 - 2r) (d + h), where h is the deviation of the heterozygote at the QTL from the midpoint between the two alternative QTL homozygotes. In both cases the observed difference between mean values of the homozygous marker genotypes in the F_2 generation provide only a minimum estimate for effects at the QTL and do not provide any information about map distance between QTL and marker. Methods for disentangling (d), (h), and (r) will be described later.

The factor (1 - 2r) also means that the quantitative effect associated with alternative marker genotypes by way of the linked QTL drops rapidly with increasing distance between marker and QTL. This will have marked effects on power (1 - β), as can be seen from the general expression relating required numbers per marker-genotype group (N) to type I error (α), type II error (β), expected difference between market genotypes (δ), and within-treatment standard deviation (σ) in a two-tailed test:

$$N = \frac{2(z_{\alpha/2} + z_{\beta})^2}{(\delta/\sigma)^2} \tag{1}$$

where $z_{\alpha/2}$ and z_{β} are the ordinates of the standard normal curve corresponding to the proportions $\alpha/2$ and β, respectively. Since, for the F_2 of a cross between inbred lines, as shown above,

$$\delta = 2(1 - 2r)d$$

N for given power will be inversely proportional to the <u>square</u> of (1 - 2r). Consequently, small changes in r can generate large increases in required N for given power. For example, for loci affecting 7-week juvenile body weight in a cross between inbred lines of chicken, the number of animals required per marker-genotype group will be N = 227, and the total size of the required F_2 population will be T = 911, if $\alpha = 0.01, \beta = 0.20$, r = 0.10, d = 40 g, and σ = 200 g, but if r = 0.2, N and T must be increased by a factor of 1.77.

Family Studies Within Segregating Populations

For most traits of interest in animal species, populations are generally polymorphic at QTL as well as at marker loci. In this case, the degree of linkage disequilibrium that can be generated by crossing populations is limited. In such populations, mapping can be based on the disequilibrium necessarily found within individual families within a single population. This disequilibrium obtains by virtue of the specific coupling relationships between marker alleles and QTL alleles in the parents of the family. Also, for the larger-sized species at least, rearing large numbers of cross-bred F_1 and F_2 offspring, unless carried out for commercial reasons as a part of normal agricultural practice, can be prohibitively costly. In contrast, for the highly productive commercial breeds, large numbers of individuals are available on the farm, and much information will often have been recorded on quantitative traits of individual animals. Thus, the additional effort required for marker-QTL linkage studies will be reduced to the task of obtaining DNA samples from suitable individuals and scoring these for the markers. Although much less formidable experimentally than a crossing experiment, within-population analyses depend on comparison of variances, rather than means, and hence will require much larger sample sizes than crosses between inbred lines.

Since mapping of a QTL with respect to a marker locus requires individuals that are heterozygous at both loci, mapping within populations requires that double heterozygotes be identified among the many homozygous or partially homozygous individuals making up the population. That is, for a diallelic marker and diallelic QTL, of the 10 possible genotypes at the marker and QTL:

MA/MA, MA/Ma, Ma/Ma
MA/mA, <u>MA/ma</u>, <u>Ma/mA</u>, MA/ma
mA/ma, mA/ma, ma/ma

Only offspring of the two underlined double heterozygous genotypes will be informative for presence of linkage.

Heterozygotes at the marker locus can be identified by scoring for the markers, but heterozygotes at the QTL can only be identified a posteriori by the demonstration of marker-associated quantitative effects in their offspring. Thus, a large proportion of the individuals evaluated for markers and quantitative traits will not be informative with respect to linkage. Furthermore, even in the (informative) double heterozygotes, the coupling relationships between marker alleles and QTL alleles will differ between different individuals (e.g. **MA/ma** versus **Ma/mA**). Consequently, if a simple pooled analysis, based on comparison of mean quantitative trait values of marker genotypes, is carried out, quantitative effects due to linked QTL over all individuals sharing a given marker genotype will tend to cancel one another. As a result, marker-associated quantitative effects may not be found, even in the presence of marker-QTL linkage.

This problem can be overcome by carrying out separate analyses among the progeny of a single individual (generally a male) found to be heterozygous at the marker loci. If the chosen parent is also heterozygous at a linked QTL, a difference is expected in the mean quantitative trait value of offspring receiving alternative marker alleles from their common parent (Fig. 2). This difference will equal $(1 - 2r)\alpha$, where α is the average effect of a gene substitution (Falconer, 1981); for codominance at the QTL this reduces to $(1 - 2r)d$. Note that this is only half the expected difference in quantitative trait value between the two offspring groups that are compared in the F_2 of a cross between inbred lines $[2(1 - 2r)d]$. Hence total number of informative progeny for equal power will have to be four times as great for an analysis of this sort as for a cross between inbred lines—and these numbers will have to be obtained within the offspring of a single individual!

Furthermore, when diallelic markers are utilized, it will not be possible, in many cases, to determine which of a sire's marker alleles were transferred to a particular offspring. For example, when both marker alleles are at equal frequencies in the population, it will be possible to identify the transferred allele for only half to three-quarters of the progeny (depending on whether dams are also scored for marker genotype). The situation is much improved when polyallelic markers are used. In

a.

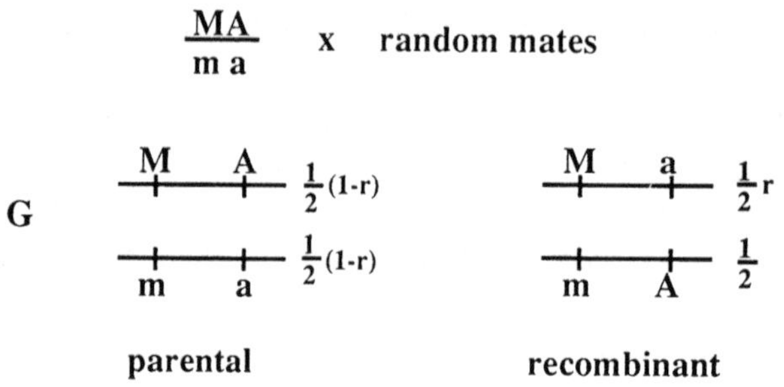

$$\frac{MA}{m\,a} \quad x \quad \text{random mates}$$

G

M A	$\frac{1}{2}(1\text{-}r)$	M a	$\frac{1}{2}r$
m a	$\frac{1}{2}(1\text{-}r)$	m A	$\frac{1}{2}$
parental		recombinant	

b.

QTL genotypes	Marker Alleles	
	M	m
AA	p(1-r)	pr
Aa	q(1-r) +pr	p(1-r) +qr
aa	qr	q(1-r)

c.

$$\overline{M} - \overline{m} = (1\text{-}2r)\,[d + (q\text{-}p)\,h]$$
$$= (1\text{-}2r)\,\alpha$$

Figure 2 Linkage between marker locus and QTL in the progeny of a sire heterozygous at both marker locus and QTL. p and q, frequency of A and a QTL alleles; M and m, respective mean quantitative value of all offspring receiving the **M** or **m** marker alleles from the heterozygous sire; α, mean effect of an allele substitution at the QTL; other symbols as in Figure 1. (a) Genotype of heterozygous sire and genotype and frequency of parental and recombinant gametes that he transmits. (b) Genotype frequencies at the QTL, within progeny receiving alternative marker alleles from the heterozygous sire. (c) Difference between mean values of the two progeny groups receiving alternative marker alleles from their heterozygous sire.

this case, each marker allele will be individually rare, and it will be possible to trace it from sire to offspring in virtually all instances (Kashi et al., 1990).

Because of the large numbers of offspring required for reasonable power, the single-individual analysis, at a theoretical level at least, is virtually limited to offspring of individual dairy cattle sires, each of which can have many hundreds of daughters by artificial insemination. Analyses of this sort were recently carried out by Geldermann et al. (1985) on three sires of the German Friesian dairy cattle breed, each having numerous offspring. In a noteworthy display of statistical audacity, Beever et al. (1990) examined marker-quantitative trait associations in 146 (!) offspring of a single sire of the Angus beef breed. In both studies significant marker-associated quantitative effects indicative of linked QTL were found.

Limitations in the number of offspring per individual sire, even under artificial insemination, mean that the power of analyses based on the offspring of a single sire only must necessarily be low. In addition, in many cases a given sire, although heterozygous at the marker locus, will be homozygous at the linked QTL—even if this QTL is segregating in the population as a whole. Thus, most single-sire analyses can be expected to be nonproductive of statistically significant marker-associated quantitative effects. Clearly, therefore, it would be useful to be able to pool data across a number of sires, so as to increase overall power of the analysis. This was first carried out by Lowry and Shultz (1959) in poultry and by Neimann-Sorensen and Robertson (1961) in dairy cattle. They reasoned that in the offspring of different sires heterozygous at the same marker allele, the difference in quantitative trait value between offspring receiving alternative marker alleles from their sire would vary. If the sire were homozygous at the QTL (**MA/mA** or **Ma/ma**), the expected difference would be zero, while if the sire were heterozygous at the QTL, the expected difference would sometimes be positive (sire genotype: **MA/ma**) and sometimes negative (sire genotype: **Ma/mA**). This would be expressed as a sire x marker interaction effect that could be detected as such in a two-way ANOVA. Results of the analysis carried out by Neimann-Sorensen and Robertson (1961) were negative, but this was attributed to lower power.

Alternatively, effects at the marker can be considered analogous to the dam-within-sire component of the familiar hierarchical ANOVA used for estimating genetic components of variance in a population having a full-sib family structure (Becker, 1967). In this case, marker-QTL linkage will be detected as a significant marker component of variance with markers nested within sires (Soller and Genizi, 1978). Analyses of this sort were recently carried out by Gonyon et al. (1987) and Haenlein et al. (1987). In these studies a number of highly significant effects were found giving strong evidence of marker-QTL linkage.

The power of experiments of this type, as a function of number of sires examined and number of daughters per sire, was considered by Neimann-Sorensen and Robertson (1961) and calculated exactly by Soller and Genizi (1978), using a rather cumbersome procedure. Recently, Weller et al. (1990), based on a procedure suggested by Neimann-Sorensen and Robertson (1961) and Geldermann (1975), introduced an approximation to the power using the chi-squared distribution. QTL having main effects (d) equal to 0.1σ will be very hard to map, even in very large experiments (5000–10,000 offspring), while QTL having main effects equal to 0.3α will be readily mapped even in small experiments (1000–2000 offspring). It is noteworthy that for any given number of daughters, greater power is achieved with fewer sires and more daughters per sire than with more sires with fewer daughters per sire. Thus, this method would be appropriate for mapping within highly improved populations that are under selection and in which large families can be obtained, e.g., the larger dairy cattle breeds in which extensive use of artificial insemination is common, providing many large half-sib families each descended from a single sire.

Crosses Between Populations that Are Segregating at the Marker Loci, but Are at or Close to Fixation for QTL

Segregating populations are often found that differ radically in quantitative value for a particular trait. These populations can be expected to differ markedly in allelic frequencies at QTL affecting the trait, approaching fixation for alternative alleles at these loci. With respect to marker loci, however, the populations may share the same marker alleles, at roughly similar frequencies. Examples of this situation would include the end-products of two-way selection experiments and commercial populations that have been selected for widely differing biological niches (e.g., extreme dog types, layer-type and broiler-type chickens, Jersey as compared to Charolais cattle). This category would also include native breeds that are resistant to specific diseases, e.g., the trypanotolerant N'Dama cattle of West Africa (Soller and Beckmann, 1987) and the tick-resistant zebu cattle of India.

In these situations, mapping cannot be carried out by the methods described in the previous section, since most individuals will be homozygous at the QTL affecting the trait. Yet, if the populations are crossed, the expected difference in quantitative trait value between homozygous marker genotypes (σ) will be proportional to the difference in allelic frequencies at the marker locus in the two populations (Soller et al., 1976). Thus if the F_2 is analyzed as a cross between two inbred lines, sample sizes required for given power [as determined by Eq. (1)] will be enormously increased when marker allele frequencies in the two populations are similar. Alternatively, the F_2 could be analyzed by the methods previously described, but this would also require large sample sizes and would not make use of the information at hand as to allelic frequencies at the QTL.

The key for F_2 analysis in this case is to use the markers to classify the offspring as to whether the marked chromosomal region derives from the "high value" population or the "low value" population. This can be done by tracing specific marker alleles from parents to F_2 offspring separately within the lineage of each individual. In this way it will be possible to pool F_2 marker genotypes having the same expected QTL composition (Beckmann and Soller, 1988). A major problem with this approach is that it is necessary to trace marker alleles from parent to offspring through two generations. For diallelic markers this will frequently not be possible. Thus, only a small proportion of F_2 progeny will be informative for marker-QTL linkage studies. In the worst case (allelic frequencies of the two marker alleles equal 0.5 in both populations), four times as many offspring overall will be required for given power as in the cross between inbred lines. For polyallelic markers, as above, each of the marker alleles can be expected to be individually rare. This will make it possible to trace specific alleles from parent to offspring much more readily. In this case, the power of such analyses will approach that of crosses between inbred lines. Additional details on analytical methods appropriate for crosses of this type can be found in Beckmann and Soller (1988).

Linkage Disequilibrium Within a Population

The majority of DNA-level genetic polymorphisms can be expected to be selectively neutral. Consequently, population-wide linkage disequilibrium between marker loci and QTL in a closed population that has not undergone a crossing episode in its recent history can only be generated by genetic drift. The degree of disequilibrium to be expected in a finite population under these assumptions was investigated by Hill and Robertson (1968) and was shown to be primarily a function of effective population size (N_e) and proportion of recombination between marker and QTL (r). Their results show that appreciable disequilibrium is to be expected when $r \leq 1/N_e$, and that the number of generations required to reach maximum disequilibrium, starting from a population in equilibrium, will be between $N_e/2$ and N_e. For many populations of agricultural animals N_e will e rather small (between 25 and 100), owing to the intense selection practiced on males. Many of these populations have also been maintained as pure breeds, closed to outside introductions, for well over 25 generations. Thus, some degree of linkage disequilibrium on a population basis could be expected for marker-QTL associations where $r \leq 0.05$. Even under such conditions, of course, not every marker will show disequilibrium with every QTL within the prescribed distance. Simulation studies in our laboratory (Z. Marko, unpublished data), however, show that under the above conditions ($25 \leq Ne \leq 100$; $0 \leq r \leq 0.05$), significant linkage disequilibrium is generated in a a large proportion of instances, and power of a test for marker-QTL linkage based on a population-wide comparison of quantitative effects associated with marker genotypes will be generally greater than 0.5, given a sufficient number of individuals scored ($1000 \leq N \leq 10000$) and depending, of course, on main effect at the QTL.

Achieving an average marker QTL distance of 2.5 cM does not require as many markers as might be expected—this would be achieved at an average marker-to-marker spacing of 10 cM, i.e., at some 300 well-spaced markers for the typical mammalian genome. Clearly, for a given quantitative trait, only a small proportion of the markers investigated would be expected to have a linked segregating QTL. For a trait of given heritability, the number of segregating QTL will be inversely proportion to the average effect at a QTL. Consequently, the overall likelihood of detecting a marker-associated effect with a given number of individuals scored is roughly independent of the number of segregating QTL. For a trait of heritability 0.25, 1000 individuals scored for markers, a marker spacing of 10 cM, and type I error of 0.01, one could expect to find at least two or three significant marker-associated effects among the entire range of markers screened. At the specified type I error, this is the number expected on chance alone, so that one should probably use a type I error of 0.001, with consequent decrease in power.

In principle, marker-associated quantitative effects due to linkage disequilibrium would be detected by a simple one-way ANOVA, comparing quantitative value of marker genotypes across the entire sampled population. In practice, many livestock populations consist of a small number of large sire half-sib families. In this case, specific marker alleles, particularly if rare, may be derived from a particular small subset of sires and hence may be associated with overall breeding value of these sires for some particular trait. This alone could generate a significant marker-genotype–associated quantitative effect. In order to avoid such confounding, it will be preferable to carry out the analyses on a hierarchical basis, with marker genotypes nested within sires, or even within sire/grandsire combinations.

Estimating Map Distance of QTL from Marker and Main Effects at the QTL

The methods described earlier can determine the presence of a QTL in linkage to a marker locus, but do not provide information as to map distance between marker and QTL. Also, because the marker-associated quantitative effects obtained in these experiments confound gene effect at the QTL and proportion of recombination between QTL and marker, they do not provide accurate estimates as to quantitative effect of the QTL locus itself. This section describes two methods that enable map distance between QTL and marker and quantitative effects at the QTL to be determined. One is based on application of maximum-likelihood methods to a single marker. The other is based on utilization of a pair of linked markers that bracket the QTL.

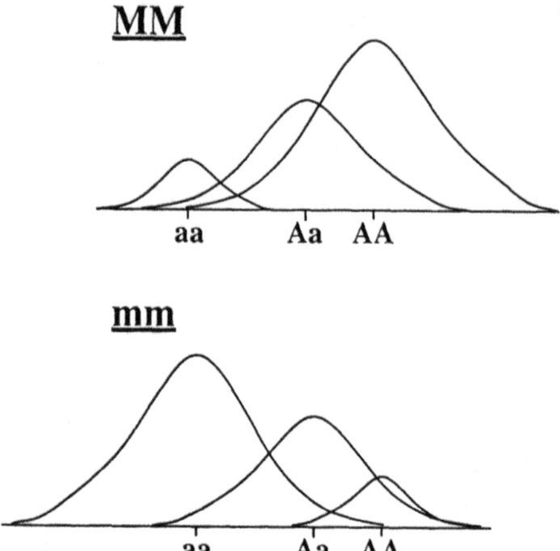

Figure 3 Distribution of quantitative value of the two homozygous marker genotypes in the F_2 of a cross between inbred lines, according to effect and frequency of genotypes at the QTL. All symbols as in Figure 2. Note opposite skewness of the two marker genotypes as a result of recombination between marker and QTL. It can also be observed that quantitative value of the heterozygote at the QTL relative to that of the main homozygous type will affect the variance within each marker genotype.

Maximum-Likelihood Estimates, with a Single Marker

As shown in Figure 1, as a result of recombination between marker and QTL each of the homozygous marker genotypes produced in the F_2 of a cross between inbred lines will consist of a mixture of genotypes at the QTL. Zhuchenko et al. (1979) pointed out that this will produce a skewed distribution within each of the marker genotypes (Fig. 3), with the degree of skewness depending on the proportion of recombination between marker and QTL. Recently Asins and Carbonnel (1988) have shown that when dominance is present a the QTL, the presence of recombinant types within the F_2 marker genotypes would also generate a difference in the within-marker-genotype variance between the two marker genotypes (see Fig. 3 for a visual demonstration). This means that in marker-QTL linkage analyses, the shape of the within-marker-genotype distributions carries information over and above that given by the observed differences in their mean value. These (marker-associated) differences in shape of the distribution are virtually inaccessible to linear AVONA, but can be exploited by moment or maximum-likelihood methods.

Along these lines, Zhuchenko et al. (1979) attempted to estimate r from the degree of skewness using the moments method. The approach, although theoretically sound, did not work out well in practice, since many estimates outside of the parameter space (i.e., negative values for r) were obtained. This is a general problem with linear-model or moment-method estimates. Maximum-likelihood methods, which can only yield estimates within the parameter space, are immune to this problem. Maximum-likelihood methods were first applied to QTL mapping by Weller (1986, 1987) in order to disentangle main effects at the QTL and proportion of recombination between marker and QTL in the F_2 of a cross between two inbred lines, enabling individual QTL to be mapped with respect to a single marker.

Obtaining maximum-likelihood estimates for situation of this sort requires maximizing the likelihood with respect to a large number of parameters: means of the three QTL genotypes r, and standard deviations within QTL genotypes. The maximum cannot be obtained analytically, and Weller (1986, 1987) used moment-method estimates for some parameters and scanned the remaining multidimensional space to find an optimum with respect to the remaining parameters. As an alternative to scanning the parameter space, it is possible to obtain the maximum of the likelihood function by use of general-purpose algorithms for finding the maximum of a function or by methods specifically for finding the maxima of likelihood functions. These include expectation-maximization (EM) (Dempster et al., 1977) and Newton-Ralphson algorithms (Dahlquist and Bjorck, 1974). Both algorithms will be more efficient than general-purpose algorithms for large samples. EM is guaranteed to converge to a maximum, although the rate of convergence can be slow. Newton-Ralphson will generally converge much more rapidly than EM and also readily provides standard errors of the estimates, but it is not guaranteed to converge, even if a maximum is present in the parameter space. Standard errors of EM estimates can also be obtained, with a little extra effort.

Linear Estimates, Using a Bracket of Markers

When, in a cross between two inbred lines, the two parent lines differ with respect to a pair of linked markers (say, **MS/MS** and **ms/ms**) bracketing a chromosomal region, the expected quantitative value of parental (**MS** and **ms**) and recombinant (**Ms** and **mS**) genotypes in the backcross or F_2 generation will differ, depending on whether a QTL (**A**) is located to either side of the bracket (**AMS** or **MSA**) or within the bracket (**MAS**). In particular, the expected ratio R of the difference between recombinant-type means (**Ms** - **mS**) to the difference between parental-type means (**MS** - ms) will be 1.0 when gene order is **AMS**, -1.0 when gene order is **MSA**, and $(r_2 - r_1)/r$ when gene order is **MAS**, where r_2, r_1, and $r = r_1 + r_2$ are the proportion of recombination between **MA**, **AS**, and **MS**, respectively (Fig. 4). Since R and r can both be calculated from the data, this enables r_1 and r_2 to be obtained for a bracketed QTL, in this way mapping the QTL with respect to **M** and **S**. This is analogous to the three-point test cross used for mapping mendelian loci.

This method was first proposed by Thoday (1961) for a backcross design, in which case

$$R = \frac{[(\mathbf{Ms/ms}) - (\mathbf{mS/ms}}{[(\mathbf{MS/ms} - (\mathbf{ms/ms})]}$$

It can readily be extended to a F_2 population. In this case,

$$R = \frac{[\{(\mathbf{Ms/MS}) + (\mathbf{Ms/Ms}) + (\mathbf{Ms/ms})\} - \{(\mathbf{mS/MS}) + (\mathbf{mS/mS}) + (\mathbf{mS/ms})\}]}{[(\mathbf{MS/MS}) - (\mathbf{ms/ms})]}$$

Application of marker brackets to data pooled over a number of sires is similarly straightforward. Note that

$$R = \frac{[(\mathbf{Ms/XX}) - (\mathbf{mS/XX})]}{[(\mathbf{MS/XX}) - (\mathbf{ms/XX})]}$$

(where **XX** is the genotype contribution of the dams) will have the same expectation independently of the distribution of QTL genotypes among the individual sires.

Statistical Techniques for Increasing the Power of Marker-QTL Linkage Experiments

Maximum-likelihood analyses and marker brackets, in addition to providing information about map distance of QTL from marker, can also be used to increase the power of the marker-QTL linkage determination itself. They do this, basically, by disentangling the effect of recombination on marker-associated quantitative differences, so that these reflect the full effect of the QTL.

Maximum Likelihood

The presence of a QTL in the vicinity of a marker will affect not only the mean value of the alternative marker genotypes in the BC or F_2 generation of a cross between inbred lines, but also the shape of the distribution, as shown in Figure 4. For this reason, maximum-likelihood methods, which can exploit this information, are able to provide greater power for detection of marker-QTL linkage, compared to linear models. Indeed, Simpson (1989), using a general-purpose algorithm for obtaining the maximum of the likelihood function, has shown that maximum-likelihood methods can markedly increase the power of linkage determinations between a single marker and a linked QTL. Maximum likelihood was particularly effective when the proportion of recombination between marker and QTL was large. Simpson (1989) analyzed the case of a set of inbred lines. Relative power of maximum-likelihood methods as compared to linear models for analysis of backcross or F_2 populations has not yet been determined, but can be expected to be equally effective when applied to these situations.

Maximum-likelihood methods can readily be applied to analysis of marker-QTL linkage in the offspring of a single sire. Here, too, recombination will produce a mixture of genotypes at the QTL within each of the daughter groups that received alternative marker alleles from their heterozygous sire (Fig. 5). The effects on within-marker-genotype distribution are less pronounced, however, than in the F_2 or BC of a cross between inbred lines, so that the contribution of maximum likelihood to power or mapping can be expected to be correspondingly less. Weller (1990) has written the likelihood equations for data pooled across sires, but actual data analysis will require maximizing with respect to sire

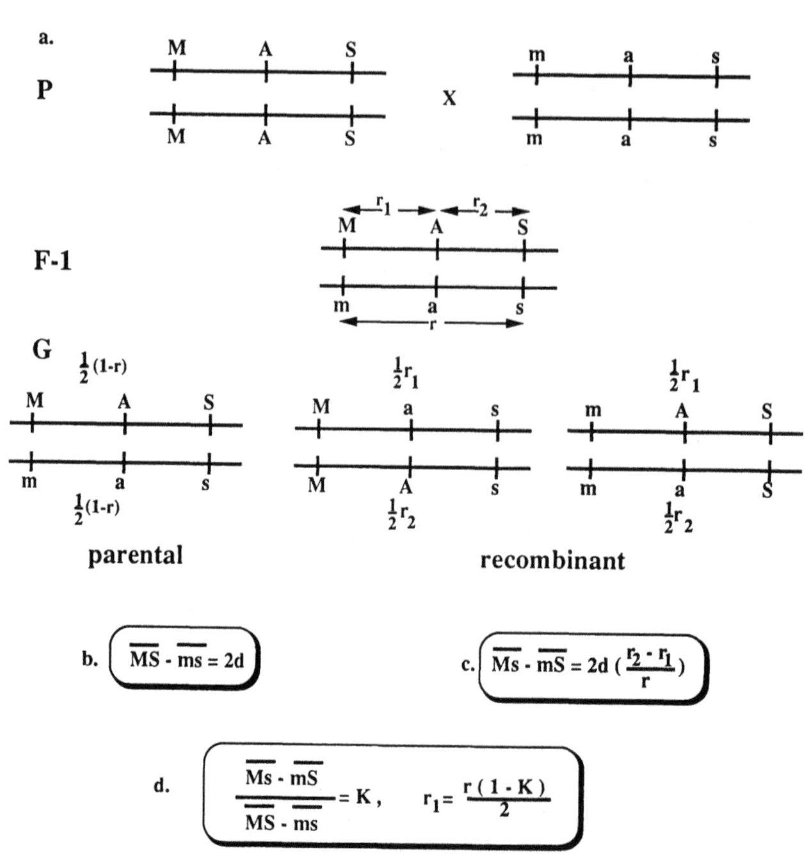

Figure 4 The use of a pair of markers to locate a QTL "bracketed" between them. **M** and **S**, marker loci; r_1, map distance between marker locus **M** and the QTL; r_2, map distance between marker locus **S** and the QTL; r, proportion of recombination between **M** and **S**; $\overline{\mathbf{MS}}$, $\overline{\mathbf{ms}}$, $\overline{\mathbf{Ms}}$ and $\overline{\mathbf{mS}}$, mean value of homozygous marker genotypes in the F_2 generation (for simplicity, heterozygous marker genotypes are not considered). Other symbols as in Figure 2. (a) Genotypes of parents, F_1 progeny, and gametes formed by F_1, showing coupling between marker alleles and QTL alleles, and frequency of parental and recombinant F_1 gametes. (b) Difference between mean values of the two homozygous parental genotypes in the F_2 generation. (c) Difference between mean values of the two homozygous recombinant genotypes in the F_2. (d) r_1 as a function of $k = [c]/[b]$ and r.

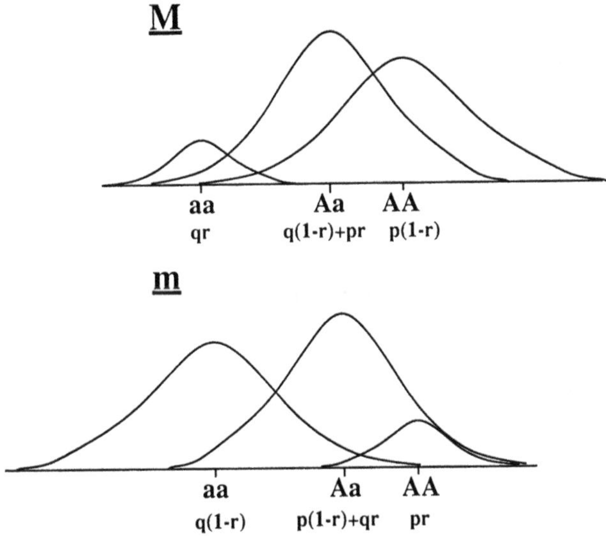

Figure 5 Distribution of quantitative trait value of the two progeny groups that received alternative marker alleles from their heterozygous sire, according to effect and frequency of genotypes at the QTL. All symbols as in Figures 2 and 4. Note opposite skewness of the two marker groups as a result of recombination between marker and QTL. Differences in within marker-group variance can also be produced as a function of dominance and allele frequencies at the QTL.

genotypes and progeny allele frequencies at the QTL, in addition to main effect at the QTL and proportion of recombination between marker and QTL. This will not be a trivial task.

Marker Brackets

Marker brackets, in addition to providing estimates of QTL location, can also disentangle the confounding of main effect and recombination found in quantitative effects associated with a single marker. The reason for this is that the difference between parental types for the marker bracket provides an estimate of the main effect of the QTL, virtually independent of recombination. In the worst-case situation (QTL midway between the two bracketing markers),

$$\text{MS/ms} - \text{ms/ms} = \left(\frac{1 - r^2}{2}\right)(d + h)$$

for the backcross design, and

$$\text{MS/MS} - \text{ms/ms} = 2d\left(\frac{1 - r^2}{2}\right)$$

for the F_2 design.

The term $r^2/2$ will generally be less than 0.1 for most mapping exercises. Consequently, when the QTL is located at or near the midpoint of the bracket, basing the analysis on parental types for the marker bracket alone, without including recombinants, can provide a marked increase in power when the QTL is located at or near the midpoint of the bracket, as compared to analysis based on either of the bracketing markers separately (Lander and Botstein, 1989; Soller, 1990a). Separating parental types and recombinants makes the parental types essentially monomorphic with respect to the QTL, so that application of maximum-likelihood methods based on distribution shape has little to offer to their analysis. The recombinant types, however, will be a mixture of QTL genotypes and hence, in addition to conveying information on the location of the QTL, will also carry information on main effect at the QTL. Lander and Botstein (1989) show that this information can also be exploited by maximum-likelihood methods (which they term "interval mapping") to provide a further (modest) increase in the statistical power of marker brackets. They indicate that interval mapping will not be able to exploit information in the recombinants when the QTL is at the midpoint between the bracketing markers (in which case the mean value of the two recombinant types will be the same). They may be underestimating the effectiveness of "interval mapping." Even when the QTL is at the midpoint, maximum likelihood should be able to utilize information conveyed by the recombinant types being a mixture of QTL genotypes, with consequent effects on the variance of recombinant compared to parental types. Interval mapping has recently been applied to

QTL mapping in a tomato backcross using DNA-level markers (Paterson et al., 1988). Lander and Botstein (1989) used EM to find the maximum of the likelihood function. To simplify matters, they assumed absence of double crossing-over and found ML estimates for the means and variances for a given value of r, and then scanned the parameter space for the r value that gave the greatest overall likelihood. Darvasi et al. (1990) used the Newton-Ralphson method in a simulation of the same experimental situation assuming the presence of double-crossing-over. By accounting for double-crossing-over, this should provide a modest increase in power relative to interval-mapping when the marker bracket is wide.

Reducing the Number of Offspring Scored for Markers by Evaluating Larger Numbers of Offspring for Quantitative Traits

Two additional statistical developments—selective genotyping and replicated progenies—are concerned with the possibility of reducing the number of offspring that need to be scored for markers by scoring increased numbers of offspring for the quantitative traits. Such methods are particularly attractive for marker-QTL mapping in dairy cattle, since quantitative data on large numbers of dairy cows are routinely collected for purposes of herd management and progeny testing. The use of replicated progenies would also be applicable to recombinant inbred lines (Soller and Beckmann, 1990).

Selective Genotyping

The first of these methods, "selective genotyping" is based on the observation by Stuber et al. (1980) that selection for quantitative traits changed marker allele frequencies in segregating populations. This suggested that a significant difference in marker allele frequencies in the selected high and low tails of an F_2 or BC population could serve as a test for marker-QTL linkage. On the basis of marker-allele frequencies, it was shown that when the proportion selected in the tails was very low (10% or less) significant savings in numbers of offspring scored for the markers could be obtained at the expense of a rather large increase in the number of offspring scored for the quantitative trait (Lebowitz et al.,

1987). Later, Lander and Botstein (1989) applied interval mapping to selected population tails, obtaining a further increase in power, over and above that obtained by considering marker allele frequencies in the selected tails. It can be shown, however (Soller, 1990b), that the increase in power obtained by Lander and Botstein results from considering quantitative value of the marker genotypes in the selected tails and will also be obtained by ANOVA without the necessity for application of maximum-likelihood methods. Selective genotyping could readily be applied, e.g., to dairy cattle data by selecting only high-and low-producing daughters of each sire for marker scoring with respect to one or two of the most important traits (say milk production and protein content). This would increase power markedly with respect to the selected traits, while not having any effect on power with respect to unselected independent traits. For example, taking the top and bottom 10% of offspring for marker analysis would require increasing the total number of daughters per sire scored for the quantitative trait by 150%., but would require scoring only 30% as many daughters for the markers. However, selective genotyping probably cannot be applied to more than two independent traits simultaneously, unless the proportion selected is very small, since this would require essentially scoring the entire (enlarged) population for the markers (Lebowitz et al., 1987).

Progeny Testing

The second method for reducing number of offspring scored for markers at the expense of increased numbers scored for the quantitative trait is to progeny-test the offspring in order to decrease the error variance of quantitative trait evaluation (Soller and Beckmann, 1990). This can be readily applied to dairy cattle data because of the fact that a small number of elite sires father most of the young progeny-tested sires of any given generation. In this design (termed a "granddaughter design"), sons, rather than daughters of a heterozygous elite sire are scored for markers and divided into two groups on the basis of the marker allele transmitted from their sire (Weller et al., 1990). Each of the sons would then be progeny-tested in order to estimate his quantitative trait value (Fig. 6). In this design the expected difference between the mean progeny tests of sons receiving alternative marker alleles from their elite grandsire would be only half that expected between daughter groups receiving alternative marker alleles from their sire. This would tend to de-

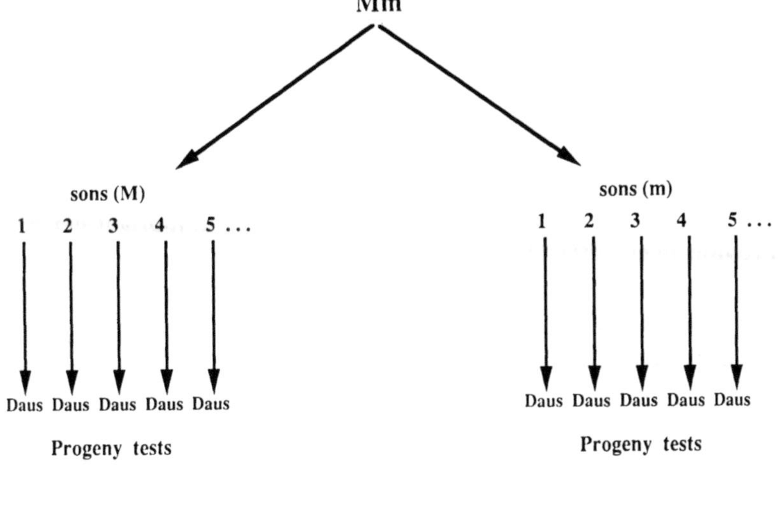

b

$\bar{M}-\bar{m} = (1-2r)\,\alpha/2$ [Only half of daughters receive M or m from elite grandsire]

Figure 6 Granddaughter design. (a) Sons of a heterozygous sire are divided into two groups on the basis of marker allele transmitted from their sire. Each son is progeny-tested to determine his quantitative trait value. (b) Expected difference in quantitative trait values between mean progeny tests of sons receiving alternative marker alleles from their heterozygous sire. Symbols as in Figure 2 and 4.

crease the power of the granddaughter design, compared to daughter designs. For low-heritability traits, however, the variance between progeny tests of sons of a sire will be much less than the variance between individual daughters, more than compensating for the above. For example, an experiment involving 20 elite sires, 50 sons per marker class per sire, and progeny test of 50 daughters per son would provide a power of 0.98 for a gene effect of d = 0.2 and heritability 0.10, and a power of 0.90 for the same gene effect and a heritability of 0.2. This would require 2000 marker assays-power for a daughter design with the same number of assays would be 0.44.

CONCLUSION

Identifying and mapping the genes we are working with as breeders is now an eminently feasible goal. The requisite populations are for the most part extant or can be readily produced. The molecular markers are there for the searching. Mapping these agriculturally important genes will provide an essential step on the way to their cloning and will allow more effective breeding manipulation of production traits and increased insight into their physiology. These goals are becoming increasingly attractive as we enter the last decade of our century.

ACKNOWLEDGMENT

This study was supported by a grant from the U.S.-Israel Agricultural Research and Development Fund (BARD). Helpful comments of J. I. Weller are acknowledged with thanks.

REFERENCES

Asins MJ, Carbonell EA. Detection of linkage between restriction fragment length polymorphism markers and quantitative traits. Theor Appl Genet 1988; 76:623-626.

Becker WA. Manual of procedures in quantitative genetics. Program in genetics. Washington State University, 1967.

Beckmann JS, Soller M. Detection of linkage between marker loci and loci affecting quantitative traits in crosses between segregating populations. Theor Appl Genet 1988; 76:228-236.

Beever JE, George PD, Fernando RL, Stormont CJ, Lewin HA. Associations between genetic markers and growth and carcass traits in a paternal half-sib family of Angus cattle. Animal Sci 1990; 68:337-344.

Dahlquist G, Bjorck A. Numerical methods. Englewood Cliffs, NJ: Prentice-Hall, 1974.

Darvasi A, Soller M. Optimum distribution of genetic markers for mapping of quantitative trait loci, 1990 (in preparation).

Darvasi A, Soller M, Weller J. Maximum likelihood mapping of quantitative trait loci with respect to genetic markers using the Newton-Ralphson algorithm, 1990 (in preparation).

Dempster AP, Laird NM, Rubin DB. Maximum likelihood from incomplete data via the EM algorithm. J Roy Stat Soc (Series B) 1977; 39:1.

Falconer DS. Introduction to quantitative genetics. New York: Longman, 1981.

Geldermann H. Investigators on inheritance of quantitative characters in animals by gene markers. I. Methods. Theor Appl Genet 1975; 46:319-330.

Geldermann H, Pieper U, Roth B. Effects of marked chromosome sections on milk performance in cattle. Theor Appl Genet 1985; 70:138-146.

Gonyon DS, Mather RE, Hines HC, Haenlein GFW, Arave CW, Gaunt SN. Associations of bovine blood and milk polymorphisms with lactation traits: Holsteins. J Dairy Sci 1987; 70:2585-2598.

Haenlein GFW, Gonyon DS, Mather RE, Hines HC. Associations of bovine blood and milk polymorphisms with lactation traits: Guernseys. J Dairy Sci 1987; 70:2599-2609.

Hill WG, Robertson A. Linkage disequilibrium in finite populations. Theor Appl Genet 1968; 38:226-231.

Kashi Y, Hallerman EM, Soller M. Marker-assisted selection of candidate sires for progeny testing programs. Animal Prod, 1990; 51:63-74.

Kerem B, Rommens, JM, Buchanan, JA, Markeiwicz D, Dox, TK, Chakravarti A, Buchwald M. Tsui L. Identification of the cystic fibrosis gene: genetic analysis. Science 1989; 245:1073-1080.

Lander ES, Botstein D. Mapping mendelian factors underlying quantitative traits using RFLP linkage maps. Genetics 1989; 121:185-199.

Lebowitz RJ, Soller M, Beckmann JS. Trait-based analyses for the detection of linkage between marker loci and quantitative trait loci in crosses between inbred lines. Theor Appl Genet 1987; 73:556-562.

Litt M, Luty JA. A hypervariable microsatellite revealed by in vitro amplification of a dinucleotide repeat within the cardiac muscle actin gene. Am J Hum Genet 1989; 44:397-401.

Lowry DC, Shultz F. Testing association of metric traits and marker genes. Ann. Hum Genet 1959; 23:83-90.

McMillan I, Robertson A. The power of methods for detection of major genes affecting quantitative characters. Heredity 1974; 32:349-356.

Nakamura Y, Leppert M, O'Connell P, Wolff R, Holm T, Culver M, Martin C, Fujumoto E, Hoff M, Kumlin E, White R. Variable number of tandem repeat (VNTR) markers for human gene mapping. Science 1987; 235:1616-1622.

Neimann-Sorensen A, Robertson A. The association between blood groups and several production characters in three Danish cattle breeds. Acta Agr Scand 1961; 11:163-196.

Paterson, AH, Lander ES, Hewitt JD, Peterson S, Lincoln SE, Tanksley SD. Resolution of quantitative traits into mendelian factors by using a complete linkage map of restriction fragment length polymorphisms. Nature 1988; 335:721-726.

Rommens, JM, Iannuzzi MC, Kerem B, Brumm ML, Melmer G, Dean M, Rozmahel R, Cole JL, Kennedy D, Hidaka N, Zsiga M, Buchwald M, Riordan JR, Tsui L, Collins FS. Identification of the cystic fibrosis gene. chromosome walking and jumping. Science 1989; 245:1059-1065.

Sax K. The association of size differences with seed-coat pattern and pigmentation in *Phaseeolus vulgaris*. Genetics 1923; 8:552-560.

Simpson SP. Detection of linkage between quantitative trait loci and restriction fragment length polymorphisms using inbred lines. Theor Appl Genet 1989; 77:815-819.

Soller M. Genetic mapping of the bovine genome using DNA-level markers with particular attention to loci affecting quantitative traits of economic importance. J Dairy Sci 1990a; 73:2628-2646.

Soller M. Relative effectiveness of interval-mapping and ANOVA for mapping of quantitative traits, 1990b (submitted).

Soller M, Beckmann JS. Toward an understanding of the genetic basis of trypantolerance in the N'Dama cattle of West Africa. Consultation report submitted to FAO, Rome, March 1987.

Soller M, Beckmann JS. Marker-based mapping of quantitative trait loci using replicated progenies. Theor Appl Genet 1990; 80:205-208.

Soller M, Genizi A. The efficiency of experimental designs for the detection of linkage between a marker locus and a locus affecting a quantitative trait in segregating populations. Biometrics 1978; 34:47-55.

Soller M, Genizi A, Brody T. On the power of experimental designs for the detection of linkage between marker loci and quantitative loci in crosses between inbred lines. Theor Appl Genet 1976; 47:35- 39.

Stuber CW, Mull RH, Goodman MM, Shaffer HE, Weir BS. Allozyme frequency changes associated with selection for increased grain yield in maize (*Zea mays* L). Genetics 1980; 95:225-236.

Suthers GK, Callen DF, Hyland VJ, Kozman HM, Baker E, Eyre H, Harper PS, Roberts SH, Hors-Cayla MC, Davies, KE, Bell MV, Sutherland GR. A new DNA marker tightly linked to the fragile X locus (FRAXA). Science 1989; 246:1298-1300.

Thoday JM. Location of polygenes. Nature 1961; 191:368-370.

Weber JL, May PE. Abundant class of human DNA polymorphisms which can be typed using the polymerase chain reaction. Am J Hum Genet 1989; 44:388-396.

Weller JI. Maximum likelihood techniques for the mapping and analysis of quantitative trait loci with the aid of genetic markers. Biometrics 1986; 42:627-640.

Weller JI, Mapping and analysis of quantitative trait loci in *Lycopersicon* (tomato) with the aid of genetic markers using approximate maximum likelihood methods. Heredity 1987; 59:413-421.

Weller JI. Experimental designs for mapping quantitative trait loci in segregating populations. In: Proceedings Fourth World Congress Genetics Applied to Livestock Production, July 1990 XIII:113-116.

Weller JI, Kashi Y, Soller M. Power of "daughter" and "granddaughter" designs for mapping of quantitative traits in dairy cattle using genetic markers. J Dairy Sci 1990; 73:2525-2532.

Zhuchenko AA, Samovol AP, Karol AB, Andryushchenko VK. Linkage between loci of quantitative characters and marker loci. II. Influence of three tomato chromosomes on variability of five quantitative characters in backcross progenies. Genetika 1979; 15:672-683 (in Soviet Genetics).

3

The Use of Reference Families for Genome Mapping in Domestic Livestock

D. J. S. Hetzel

CSIRO, Tropical Cattle Research Centre
Rockhampton, Queensland, Australia

INTRODUCTION

The construction and use of genetic maps has been greatly facilitated in recent years by technical developments in molecular biology. As a result, mapping of the entire genomes of prokaryotes as well as higher organisms such as humans is underway. Although gene mapping in domestic livestock has advanced slowly to date, the potential returns from exploiting such genetic information are now being realized. Consequently, there is considerable interest in the identification and molecular characterization of genes that are either directly related to economically important traits in livestock or that can be manipulated to further improve productivity.

A well-defined genetic map is a powerful tool for identifying potentially important genes, using a procedure known as "reverse genetics" (1). Initially, a linkage study is carried out using a panel of genetic markers, ideally selected from a genetic map on the basis of expected distribution. Once even loose linkage is detected, tightly linked markers can be defined by screening nearby markers, again selected from a genetic map. It therefore follows that the construction of primary genetic maps for livestock species such as cattle, sheep, pigs, goats, and poultry is urgently needed. Although the density of genetic maps for livestock is considerably lower than for other mammals such as human and mouse (2), the prospects for rapid progress are good.

Genetic maps are constructed by a combination of genetic and physical techniques (3). While physical mapping methods such as in situ hybridization, the use of somatic cell hybrid panels, and pulsed-field gel electrophoresis can localize genes to a specific chromosomal region, the genetic associations of genes can only be measured by the frequency with which they are coinherited. Thus genetic mapping involves an analysis of segregation within families. An efficient strategy to analyze the degree of cosegregation of genes is to focus studies on a set of reference families. Such families can be used to map polymorphic markers for use in subsequent linkage studies (4), as well as to identify homologous chromosomal segments between organisms, i.e., a comparative mapping approach similar in principal to the physical comparisons reported in Ref. 5.

This chapter will discuss the use of reference families in genome linkage mapping. The value of making use of a limited number of efficiently structured families will be emphasized, as will their role in stimulating communication and collaboration between researchers—an essential prerequisite for mapping animal genomes in a reasonable time span.

DEFINITION OF REFERENCE FAMILIES

Reference families are those used as a common resource by research groups for mapping genes or genetic markers by linkage using restriction fragment length polymorphism (RFLP) markers or other types of markers. The family structures should therefore be efficient for mapping a wide variety of genes or markers from both highly and lowly polymorphic systems. *Resource families* are those in which specific genes of biological or economic importance or traits controlled by a number of genes are segregating or likely to be segregating. Where there is appropriate phenotypic or performance information on family members, the gene(s) may be mapped using the reverse genetics approach. There is no reason why resource families cannot be used as reference families. Indeed, this approach is most efficient since at the same time as accumulating linkage information to construct genetic maps, markers can be screened for linkage with genes affecting specific traits. However, it should be remembered that the most appropriate family structures for reference and resource families will often differ. For example, families with identifiable single genes segregating, e.g., genetic diseases, will frequently involve the mating of related individuals. However, the most efficient families for general linkage mapping are based on the mating of outbred individuals. Furthermore, it is often difficult to generate large full-sib families in species such as cattle and sheep at the same time as collecting meaningful phenotypic information. Nevertheless, it is highly desirable for families to serve both purposes. This discussion will be restricted to the use of reference families for general linkage mapping leading to the construction of genetic maps.

CENTRE D'ETUDE DU POLYMORPHISME HUMAIN

The Centre d'Étude du Polymorphisme Humain (CEPH) was established in Paris in 1984 to facilitate the development of a linkage map of the human genome (19). CEPH provides high-quality DNA to collaborating investigators who map genes or markers by linkage, using their own probes within the family panel. All results are contributed to a central database, which is made available to all collaborators.

As the RFLP studies are conducted in a common set of families, the results can be readily pooled, Collaborators must agree to carry out the mapping in all informative families within the panel. Although individual collaborators are free to publish their own data, consortium maps of individual chromosomes making use of all available information are now being constructed.

Until recently, the CEPH family panel consisted of 40 full-sib families, 29 of which included all four grandparents (19). The mean sibship size was 8.2. However, in order to construct a 1 cM (1 cM = 1 unit of genetic recombination) genetic map, 20 more families have been added.

One of the features of the CEPH program is that it has provided a forum for other forms of collaboration and information exchange.

Given that animal genome mapping is in its infancy, collaboration will be important determinant of how rapidly genetic maps are built up and utilized. Consequently, the use of animal reference families will likely have beneficial spinoffs.

GENETIC LINKAGE

Genetic linkage can be defined as the cosegregation of pairs of genes due to their location on the same chromosome. During meiosis, crossing over or recombination (Θ) is roughly proportional to the distance between two loci on the chromosome; i.e., a crossover rarely occurs between two loci that are very close to each other. Thus the distance between two loci can be estimated from the proportion of recombination events during meiosis as observed in the genotypes of offspring, given parental genotypes. Consequently linkage analysis requires observations on groups of related individuals in which the inheritance of genes can be traced. A detailed treatment of linkage is given in Ref. 6.

Only in some matings can recombinant and nonrecombinant gene combinations be distinguished. For a mating to be informative for linkage analysis of two genes, at least one of the parents must be a double heterozygote for autosomal loci or the mother should be doubly heterozygous for X-linked loci. Therefore, the probability that an individual will be heterozygous for any gene or genetic marker chosen at random is an important determinant of whether matings involving the individuals will be useful for linkage analysis. It follows that matings between individuals that are heterozygous at a large proportion of gene loci will be highly efficient for linkage studies. The principle behind reference families in animals must be that the matings are highly likely to be informative with respect to any given pair of genes/genetic markers.

Family structures in domestic animals are usually closely controlled. It is common for males to be naturally mated with a number of females that subsequently bear one or more offspring. However, when artificial insemination is used, the number of offspring from a single male can reach many thousands. Multiple ovulation and embryo transfer techniques can be used to increase the reproductive rate of females. Nevertheless there are two basic family structures which, when present in combination, will constitute an extended pedigree (Fig. 1). The full-sib family consists of offspring from a single male (sire) and female (dam) parent, while offspring in a half-sib family have only one common parent, most frequently being progeny of multiple dams mated to a single sire. With multiparous animals such as pigs, poultry, and some breeds of sheep and goats, both full-and half-sib families are generated simultaneously. However, for uniparous species, e.g., cattle and most sheep and goats, sizable full-sib families are not produced over a reasonable time period unless technologies such as multiple ovulation and embryo transfer are used.

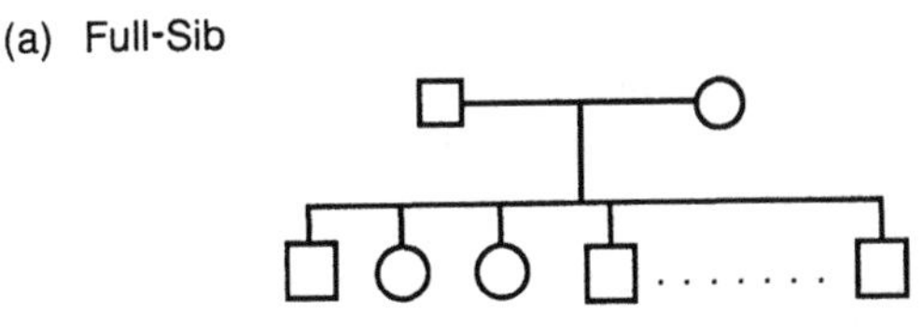

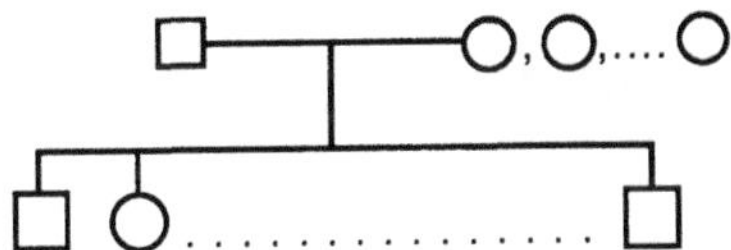

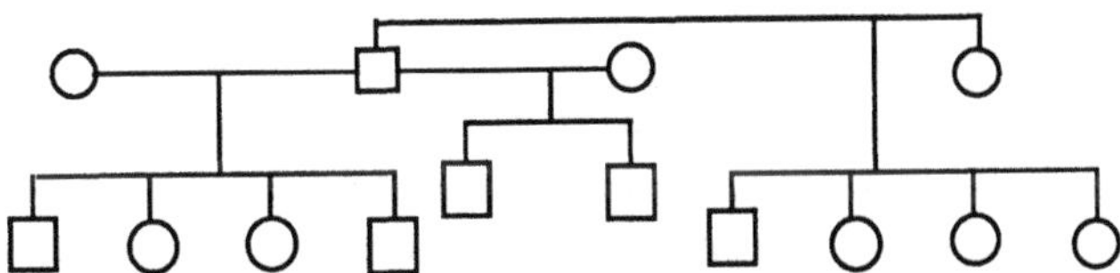

Figure 1 Examples of full-sib, half-sib, and extended families (□ = male, ○ = female).

Although extended pedigrees over multiple generations can be readily built up in animals, the discussion of reference families will be limited to two-or three-generation pedigrees.

RESTRICTION FRAGMENT LENGTH POLYMORPHISM

The number of loci that could be used for linkage analysis was rather limited until a new class of highly abundant genetic markers was discovered, namely RFLPs (7). Individual variability of DNA sequence is detected by the use of restriction endonucleases, which cut genomic DNA into small fragments according to specific recognition sequences. Point mutations, insertions, deletions, translocations, and duplications can be detected if the mutations result in different in different size fragments when cut with a specific restriction endonuclease and if a probe containing a sequence from the DNA fragment is available. Although new classes of DNA sequence polymorphisms have recently been reported (8,9), RFLPs will be the basis of primary linkage maps for domestic animals. Since DNA is relatively stable, it can be readily stored and transported, making it suitable for reference family material. Other genetic markers such as protein polymorphisms will still be valuable additions to genetic maps. However, they do not lend themselves to use on a centrally coordinated panel of reference families.

STRUCTURE OF ANIMAL REFERENCE FAMILIES

In discussing the structure of reference families for general linkage mapping, a number of issues must be addressed.

Number of Families

The number of families required to construct an RFLP linkage map will depend on the density of map required, degree of informativeness of the families, and family size. The immediate goal should be to generate primary linkage maps with a minimum spacing of 20 cM between genes/markers. From standard logarithm of odds (lod) ratio score tables, it can be calculated that a minimum of 64 informative meiotic events are required to confirm linkage of less than 20 recombination units. If the

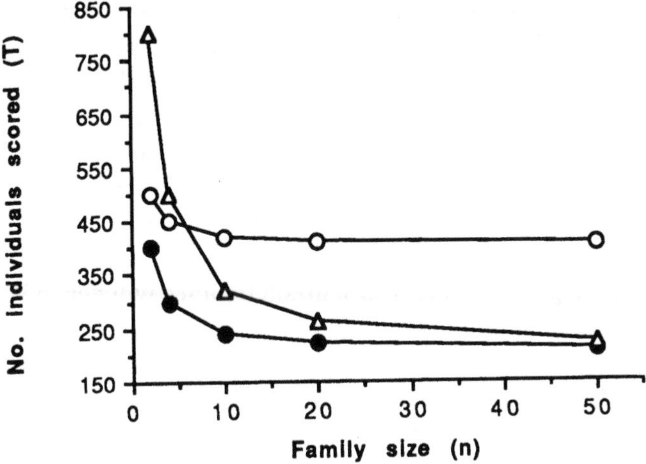

Figure 2 The total number of individuals scored to analyze 200 meiotic events from two-generation full-sib (●), half-sib (○), and three-generation full-sib with grandparents (△) families.

probability of a parent being a double heterozygote is 1 in 16, e.g., for lowly polymorphic RFLP systems, then a total of 1024 meioses will need to be scored to exclude linkage less than 20%. If, on the other hand, the frequency of heterozygous individuals exceeds 50%, such as for highly polymorphic RFLP systems including minisatellites (10), microsatellites (11), and some major satellite markers (12), then less than 256 informative meioses are required. Clearly there is an incentive to choose polymorphic parents and to concentrate mapping on polymorphic systems. All else being equal, larger families are more efficient, since the amount of information per individual (including parents) scored increases with the family size, particularly in full-sib families (see Fig. 2).

Family Type

Considering the use of full-sib, half-sib, or extended families in RFLP linkage mapping, the amount of information per individual scored for a genetic marker is maximized in full-sib families of known haplotype.

This arises through minimization of the number of parents to be typed. It has been suggested (13) that the most efficient structure for linkage mapping is to use complete three-generation pedigrees, which include the four grandparents so that linkage phase in the parents can be unequivocally determined. However, where the number of full-sib offspring is large, e.g., more than 10, phase can usually be worked out, so the grandparents are not essential. However, they do provide an internal check for consistency of segregation.

The relative advantage of full-sib over half-sib families will depend on family size. The total number of individuals scored (T) for a given number of meioses is f(2n + 1), f(n + 2), and f(n + 6) for half-sib, full-sib and full-sib + grandparent families, respectively, where f = number of families and n = number of offspring per family. Assuming the same probability of informative meioses in each case, the relative advantage of full-sib families increases as family size increases (Fig. 2). However, if family size is low and grandparents are included, the efficiency of half-sib families is improved although phase will be unknown in the latter case. Assuming full-sib family sizes over 10, the efficiency of full-sib groups for linkage mapping is nearly double that for half-sib groups. Other criteria for deciding on family structure include cost of RFLP scoring versus the cost of generating full-sib groups, especially if artificial breeding techniques are required. In general, since the repeated use of families for linkage studies will multiply any differences in efficiency many times over, it will be advisable to make use of large full-sib families, provided there is no difference in expected informativeness.

Parental Genotypes

As discussed earlier, the efficiency of reference families for general linkage mapping increases rapidly as the expected heterozygosity of parents increases. It is assumed that families will be used not only to map polymorphic markers, but also to locate genes mapped in either organisms, e.g., human and mouse, to build up comparative maps. RFLPs associated with most coding sequences exhibit a low degree of polymorphism (14) although new techniques promise to locate additional variability (8,9). However, for the immediate future, parents in reference families should be highly heterozygous at any given locus.

There are a number of approaches to maximizing parental heterozygosity.

1. Use can be made of F_1 (or subsequent generation) sires (or dams) resulting from crosses between genetically distant breeds. In cattle, the widest crosses possible that result in fertile first-cross progeny are between *Bos indicus*, *Bos taurus*, or sanga breeds. Other wide crosses are possible between dairy, beef, or draught breeds. Wide crosses are possible in sheep and goats (fiber, meat, milk breeds), chickens (meat, layer, exotic breeds/strains), and pigs (domesticated, wild breeds). Segregation can be followed in backcross or F_2 progeny. Since RFLPs segregate as codominant systems, F_2 crosses are more informative than backcrosses (6).
2. Crosses between highly inbred or highly selected lines can also be highly heterozygous. However, highly inbred lines are only available for chickens and to a much lesser extent for cattle and pigs. Closed lines that have been subjected to intense selection for one or several traits will be homozygous for many genes affecting the trait, particularly if the population size is small. In general, however, crosses between selected lines would better serve as resource families for mapping genes affecting the trait under selection.
3. Individuals from composite or synthetic breeds/strains that are based on three or more diverse breeds are also likely to be more polymorphic than purebred individuals, provided little selection or inbreeding has occurred after formation.

MANAGEMENT OF ANIMAL REFERENCE FAMILIES

The use of a common set of reference families to facilitate the construction of primary linkage maps for livestock species will require considerable organization. Because of the number of domestic livestock species that would benefit from reference families, it will be difficult, though desirable, to establish a single center for the maintenance, processing, and distribution activities. Coordination at an international level is essential, to avoid duplication of effort. Several specific issues warrant discussion.

Form of DNA Storage

The basic requirement from reference families is a continued supply of DNA for collaborators to use in linkage studies. Alternative forms of DNA storage are purified DNA, tissues such as muscle or liver, whole blood, white blood cells, or cell lines. It is difficult to predict in advance how much DNA will be required from the families over the next decade(s). Although new techniques for genetic analysis require only small quantities of DNA (15), it would be preferable to have infinite rather than finite quantities of DNA from family members. CEPH has established lymphoblastoid cell lines for all members of the family panel by transforming cells with the Epstein-Barr virus (EBV) (16). However, it appears that cells from domestic animals may be refractory to EBV. Although other viruses, e.g., *Theileria*, will transform bovine cells, gross chromosomal rearrangements do occur (H. Ansari, personal communication) which may make them unsuitable for linkage mapping. There are published methods for short-term culture of T lymphocytes, e.g., using T-cell growth factor (17). However, the methods are labor intensive and expensive.

Unless very large quantities (e.g., 100 mg) of DNA can be collected from individual animals, there is a danger of running out in the future. Thus it is highly desirable for a cost-effective method for permanent animal cell lines to be developed.

Central Database

An important feature of reference families is the contribution of all data derived from the families to a central database. Under the CEPH system, the database is available to all collaborators and it is monitored regularly to assess progress on the linkage map. CEPH also develops and maintains computer programs for data preparation and linkage analysis for collaborators. A similar system should be adopted for animal reference families.

EXISTING ANIMAL REFERENCE FAMILIES

Although the concept of animal reference families has been under discussion for several years, only a limited number of projects are underway to develop such a resource. For example, in pigs, crosses between several Chinese breeds (Meishan, Fengging, and Ming) and the Yorkshire breed are being made in the United States (L.R. Schook, personal communication), while in Sweden, crosses between the wild pig and the Swedish Yorkshire breed are planned (L. Andersson, personal communication). In both cases, data on a wide range of production traits are being collected on F_1, F_2, and some backcross generations in order to map genes for these traits; i.e., the families will be used as both reference families and resource families.

In cattle, the generation and sampling of families based on crosses between diverse breeds in progress in Australia, Kenya, and the United States. The only reference families currently in use are those established in Australia, and these are being shared on a limited basis by CSIRO, Rockhampton. DNA stocks and purified lymphocytes from 164 animals in 16 two-or three-generation full-sib families have been collected and stored. Parental genotypes are *Bos taurus* x *Bos indicus* or *Bos indicus* strain cross. The number of offspring per sire ranges from 10 to 44 since many of the full-sib families are part of extended families. Several examples of the families are given in Figure 3.

CONCLUDING REMARKS

Given that gene mapping in livestock is in its infancy, reference families will be especially valuable in rapidly constructing primary linkage maps. It is highly likely that the use of a common resource for linkage mapping will lead to greater collaboration and communication at both national and international levels. Although there is general agreement on the desirability for such a resource, the infrastructure and management of reference families should not be underestimated. Although new techniques for producing linkage maps are being developed (18), it can be expected that reference families will play an important role over the next decade. Accordingly, the structure and management of the family material should meet high standards of efficiency.

(a) **Reference Family 02**

Sire Genotype - F 1 Simmental x Brahman
Dam Genotype - F 1 Brahman x Hereford-Shorthorn
Number Full Sib Offspring - 6
Number of Grandparents - 4
Related Families - 03

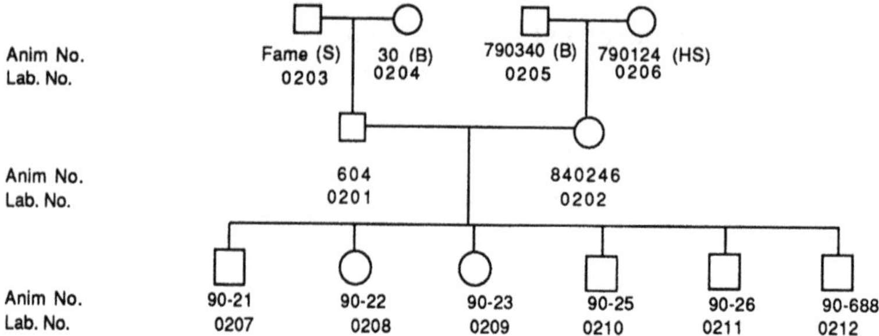

(b) **Reference Family 07**

Sire Genotype - Australian Friesan x Sahiwal
Dam Genotype - Australian Friesan x Sahiwal
Number Full Sib Offspring - 15
Number of Grandparents - 1
Related Families - 05,13,17

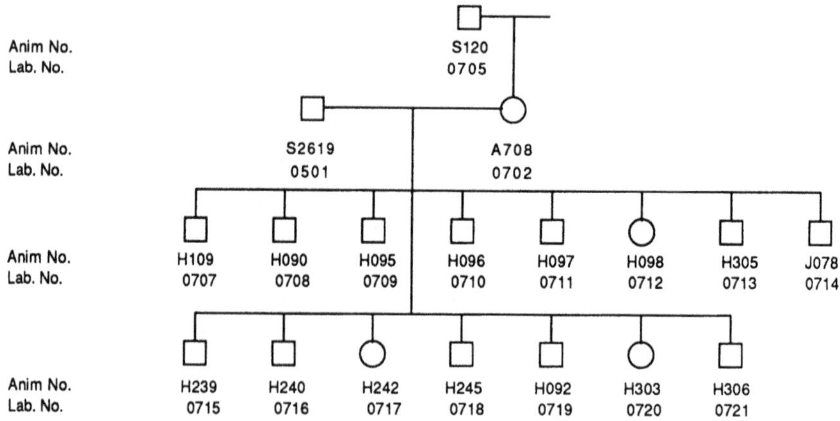

Figure 3 Examples of cattle reference families in use from CSIRO, Rockhampton, Australia.

REFERENCES

1. Orkin S. Reverse genetics and human disease. Cell 1986; 47:845-850.
2. O'Brien SJ, Ed. Genetic maps (5th ed). New York: Cold Spring Harbor Laboratory Press, 1990 Book 4. Nonhuman vertebrates.
3. Fries R, Ruddle FH. Gene mapping in domesticated animals. In: St John J, ed., 10th Beltsville Symposium on Agricultural Research. Biotechnology for solving agricultural problems. Vol X. 1986: 19-37.
4. Hetzel DJS, Fries R. Reverse genetics—a molecular approach to animal breeding Proc Aust Assoc Anim Breed Genet 1988; 7:21-31.
5. Womack JE, Moll Y. Gene map of the cow: conservation of linkage with mouse and man. J. Hered 1986; 77:2-7.
6. Ott J. Analysis of human genetic linkage. Baltimore: Johns Hopkins University Press, 1985.
7. Botstein D, White RL, Scholnick M, Davis RW. Construction of a genetic linkage map in man using restriction fragment length polymorphisms. Am J Hum Genet 1980; 32:314-331.
8. Uifterlinden AG, Slagboom PE, Knook RL, Vijg J. Two dimensional DNA fingerprinting of human individuals. Proc Natl Acad Sci USA 1989; 86:2742-2746.
9. Htun H, Dahlberg JE. Topology and formation of triple stranded h-DNA. Science 1989; 243:1571-1576.
10. Jeffreys AJ, Wilson V, Thein SL. Hypervariable "minisatellite" regions in human DNA. Nature 1985; 314:67-73.
11. Weber JL, May PE. abundant class of human DNA polymorphisms which can be typed using the polymerase chain reaction. Am J Hum Genet 1989; 44:388-396.
12. Fowler C, Drinkwater RD, Skinner J, Burgoyne L. Human satellite-III DNA: an example of a macrosatellite polymorphism. Hum Genet 1988; 79:265-272.
13. White R, Leppert M, Bishop T, Barker D, Berkowitz J, Brown C, Callahan P, Holm T, Jerominski L. Construction of linkage maps with DNA markers for human chromosomes. Nature 1985; 313:101-105.
14. Pearson PL, Kidd KK, Willard HF. Report of the committee on human gene mapping by recombinant DNA techniques. Cytogenet Cell Genet 1987; 46:390-566.
15. Saiki RK, Gelfand DH, Stoffel S, Scharf SJ, Higuchi R, Horn GT, Mullis KB, Erlich HA. Primer directed enzymatic amplification of DNA with a thermostable DNA polymerase. Science; 239:487-491.

16. Anderson MA, Gusella JF. Use of cyclosporin A in establishing Epstein-Barr virus-transformed human lymphoblastoid cell lines. In Vitro 1984; 20:856-858.
17. Morgan DA, Ruscetti FW, Gallow RC. Selective in vitro growth of T lymphocytes from normal human bone marrows. Science 1976; 193:1007-1008.
18. Li H, Gyllensten UB, Cui X, Saiki RK, Erlich HA, Arnheim N. Amplification and analysis of DNA sequences in single human sperm and diploid cells. Nature 1988; 335:414-417.
19. Dausset J, Cann H, Cohen, D, Lathrop M, Lalouel, J-M, White R. Centre d'Etude du Polymorphisme Humain (CEPH): Collaborative genetic mapping of the human genome. Genomics 1990; 6:575-577.

4

From Linked Marker to Disease Gene

Current Approaches

David L. Nelson

Institute for Molecular Genetics
Baylor College of Medicine
Houston, Texas

INTRODUCTION

The recent identification of the cystic fibrosis (CF) gene (1a–c) has further demonstrated the validity of what has become known as the "reverse genetics" (2) approach to human genetic disease. While other disease genes have been identified without prior knowledge of the protein products they encode [Duchenne muscular dystrophy (3), chronic granulomatous disease (4), retinoblastoma (5), Wilms' tumor (6)], the CF gene was the first to be identified without the aid of chromosomal deletions or translocations. The methods used to define the CF gene form the basis for similar gene hunts for the several thousand other human disease-related genes where linked markers provide the only starting point. This chapter will describe these current methods with reference to their application to a number of gene searches. It will also describe the approaches offered by the human genome initiative, pointing out the distinct advantages of a "genome approach" to human genetic disease.

REVERSE GENETICS

Reverse genetics has been defined as the use of linked markers to describe the location of a gene involved in a disease phenotype, followed by efforts to identify and isolate the gene using this starting information. Using this approach, a large number of disease genes have already been localized to chromosomal regions (7–12). The journey from the linked marker to the gene involved has been proven more difficult than anticipated. The identification of linked markers has been reasonably simple, given a well-defined disease phenotype and pedigrees with sufficient information content. Making the step from a linked marker (which can often be located several million base pairs of DNA distant from the disease gene) to the involved gene has proven to be a major undertaking. New methods and materials, described below, should accelerate the pace of identification of human disease genes from linked DNA markers.

LINKED MARKERS

It is not within the scope of this chapter to discuss the identification of linked restriction fragment length polymorphic markers for disease gene localization. This is discussed at length elsewhere in this volume. Linked markers provide the starting point for the identification and isolation of disease loci. It is clear from the state of the reverse genetics field that the identification of a linked marker has become fairly routine as additional highly informative markers are localized to genomic regions. Indeed, numerous disease gene localizations have been made (7–12), but only a few disease genes have been identified at the coding sequence level, and of these, all but CF required special insight from aberrant chromosomes in rare patients. At the time of the identification of a marker linked to Huntington's disease (HD—1983) (13), it was generally felt that finding the genetic location of the disease gene would constitute a major effort. By far the major impediment to the reverse genetics approach for identification of disease genes is the difficulty in obtaining cloned DNA from specific genomic regions.

AFTER A LINKED MARKER IS IDENTIFIED

Linked markers are typically several million base pairs of DNA distant from the disease gene, and the direction from the linked marker to the disease gene on the chromosome is usually unclear. In order to bracket the area of the chromosome to be searched, it is necessary to identify a flanking polymorphic marker. In some cases, this has been relatively straightforward, while in others it has proven to be quite a challenge. In the CF case, a flanking probe (J3.11) was found very quickly to augment the initial linkage to the *met* oncogene, the closest marker (14). This probe was derived from preexisting markers known to map to chromosome 7 and turned out to be approximately 1600 kb from *met*. While this limited the search to only 0.5% of the genome, it was still 4 years before the gene was identified. In the case of Huntington's disease, it is as yet unclear whether nor not markers flanking the HD gene have been identified. With the location of HD at the extreme terminal region of the short arm of chromosome 4 (15), most chromosome 4 probes have been proximal to the original linked marker, G8. Numerous probes in the 10,000-kb region distal to G8 have been isolated; however, none has clearly demonstrated genetic evidence of a flanking probe. This has been primarily a function of the lack of recombination in this telomeric location, as only a handful of crossover events have occurred. However, the HD gene has been flanked by proximal markers and the end of the chromosome. The recent identification of telomere containing yeast artificial chromosome clones for the short arm of chromosome 4 allows the conclusion that flanking probes now exist (16).

Once a region can be clearly defined by the isolation of bracketing probes, the distance between markers and the disease gene can be estimated. As pedigree analysis and linkage mapping is the method used to define the linkage initially, this is typically employed to provide an estimate of the distance. Flanking markers can refine the distance measurements. However, as markers are found that show very tight linkage with the disease, this method of measuring distance becomes less useful. Linkage mapping requires the identification of recombinant chromosomes, produced by crossovers between the marker and the disease. The frequency of such events for closely linked markers is small (by definition). A marker at one centimorgan (cM) shows recombination in 1% of meioses. With the added complication of the quality of the information content of the alleles segregating in a disease pedigree, the number of meioses required to accurately measure distances below 1-2 cM by linkage methods is quite large. Therefore, below this distance, it is very difficult to make use of pedigree analysis for mapping.

At close genetic distances (0.1-1 cM), it is possible to take advantage of historical crossovers (or more precisely, lack of crossovers) between linked markers and disease genes. This phenomenon is referred to as linkage disequilibrium, where markers are found out of balance (disequilibrium) with one another based on the frequencies found in the general population (17). Typically, the amount of disequilibrium is measured with regard to the diseased chromosome, and loci out of balance demonstrate the alleles on which the disease mutation arose. Linkage disequilibrium can provide an estimate of the proximity of a disease gene when most disease chromosomes demonstrate identical alleles. The amount of disequilibrium at a locus, when compared to other linked loci, can allow the determination of the distance of the locus relative to the other loci. This method is most useful for gene mutations thought to have arisen once (or only a few times), and recently, so that the background alleles remain fixed.

Closely linked markers, those within 1-2cM, can be at large physical distances. An average metric between physical and genetic distances in humans sets 1 cM equivalent to 1 million base pairs of DNA (Mb). This metric varies considerably from region to region and also between male and female meioses in the same region (18). When closely linked markers exist (as with *met* and J3.11 in the CF case), it is possible to use pulsed-field electrophoresis to derive a map of the region (19). Again, however, the development of physical maps is limited by the availability of probes that identify restriction fragments.

ISOLATION OF REGIONALLY LOCALIZED DNAs

Methods for the collection and identification of regionally localized DNAs are currently the major limitation in the reverse genetics approach to disease gene isolation. The identification of an initial linked marker begins the search for additional DNA fragments in the same portion of the chromosome that will serve as polymorphic markers and probes for physical mapping as well as for the identification of potential expressed sequences. A number of techniques exist that allow the isolation of nearby sequences. In the following discussion, these methods are divided into those that make use of preexisting DNA fragments for the isolation of nearby DNAs and those that attempt to isolate DNA fragments specifically from chromosomal regions.

REGION-SPECIFIC DNA ISOLATION

Methods for the isolation of chromosome or region-specific isolation of DNA fragments are numerous. Each takes advantage of special materials or instruments. The original method for region-specific DNA isolation is still widely used. This method, developed by James Gusella and David Housman in the late 1970s, takes advantage of somatic cell hybrids retaining small amounts of the human genome in rodent cell backgrounds (20). Such hybrid cells have been used extensively for the assignment of human markers to specific chromosomes (21). Prior to the development of recombinant DNA methods, human-specific isozyme markers were assigned by testing various hybrid cells containing differing complements of human chromosomes. With the ability to detect human DNA fragments with Southern filters (22), DNA hybridization has been widely used for chromosome assignments of genes and random DNA fragments. Gusella and Housman made use of recombinant DNA libraries derived from hybrid cells to obtain clones of human DNA from the human chromosome(s) in the hybrid. This provides a means for the isolation of chromosome-specific DNAs. Crucial to the method was a test that would allow identification of clones containing human DNA in the background of a library derived largely from rodent sequences. The recognition that human-specific repetitive sequences could be utilized for this discrimination was seminal. This method has been widely applied to a variety of systems, most notably to DNA transfection of human genes into rodent cells and the subsequent isolation of the selected gene by repetitive sequence tags (23). A number of repetitive elements have been described in the human genome; however, the most abundant is the *Alu* repeat. *Alu* repeats are found on average every 4 kilobase pairs, allowing the identification of the majority of human clones in a library of 18–20 kb inserts (~95%) (24). Repeats homologous to the human *Alu* sequences are found in rodent genomes; however, discrimination between human and rodent repeats can be accomplished by hybridization to total human DNA probes under appropriately stringent conditions.

By taking advantage of somatic cell hybrid techniques, it is possible to limit the human clones identified to quite restricted human regions. A method for the production of somatic cell hybrids containing small chromosome fragments, developed originally by Goss and Harris (25), has proven useful in this regard. The method makes use of the effects of X-irradiation on chromosomes, allowing fragmentation of chromosomes from donor cells (while also killing the donor cell) followed by fusion and selection. In this way, hybrid cells can be found that retain one or few fragments of a chromosome (it is usually advisable to begin with a reduced chromosome hybrid as a donor cell). Fragment sizes can be adjusted by the radiation dose, ranging in size from 0.5 to several Mb. After identification of cell lines with the desired region, libraries can be prepared and human clones identified. In this way, clones can be found that are derived from the desired region. It is not a trivial matter to prepare hybrid cell lines with these characteristics. The screening of hundreds of hybrid cell lines may be required to find a hybrid with a single, small fragment surrounding a marker of interest. In addition, such hybrids are not necessarily stable. The prior insertion of a selectable marker into the region desired alleviates the stability problem, but the identification of such an insertion is difficult (26).

The smaller the region in a somatic cell hybrid, the more focused a library can be for a specific region. However, small regions require bigger libraries and more screening for human clones as the percentage of clones that is human decreases. As the hybrid cell line becomes more valuable for isolating nearby fragments, it becomes more difficult to use. To circumvent the production of a recombinant library from hybrid cell lines, Nelson made use of the polymerase chain reaction (PCR) (27) to amplify human DNA specifically. Using the observation that sufficient sequence variation exists in the *Alu* repeat for differential hybridization to be possible, he designed oligonucleotides that would allow amplification of DNA between *Alu* repeats in human, but not rodent DNA (28). The method relies on two *Alu* repeats in close proximity (<4 kb) providing a template for amplification (see Fig. 1). There is variation among different oligonucleotide primers, but *Alu* PCR provides an amplified fragment roughly every 100 kb from hybrid cell DNA. The method is especially useful for hybrids with small amounts of human DNA, as it provides a means for isolating region-specific DNAs without the production and screening of recombinant libraries. In addition, by examination of the products from the *Alu* PCR, it is possible to characterize the sequences in the hybrid cell line and to compare different hybrids for possible overlapping segments. Finally, the products of *Alu* PCR can be used as probes directly and for the identification of larger cloned DNAs in libraries derived from total human DNA, allowing somatic cell hybrids to identify region specific clones rather than to produce them.

The other major method for identification of chromosome-specific DNA probes is metaphase chromosome sorting. This technology is based on the fluorescence-activated cell sorter (FACS), which was modified to also allow sorting of metaphase (condensed) chromosomes when appropriate fluorescent dyes were incorporated (29). Since sorting is time and labor intensive, and since very large numbers of chromosomes must be sorted to achieve a significant quantity of DNA, recombinant libraries from sorted chromosome material have been the means of propagating this material. Such libraries for specific human chromosomes have been produced at the Los Alamos and Lawrence Livermore National Laboratories and are readily available from ATCC. For the identification of chromosome specific DNA fragments, sorted chromosome libraries are very useful. However, for regions of less than whole chromosome size, large numbers of clones must be screened to find the small percentage that maps to the desired region. Several hundred clones from sorted libraries of chromosomes 4 and 7 have been regionally assigned during the pursuits of HD and CF, respectively. This is time-consuming, as subregional localization is required of each clone individually.

Another method that allows the isolation of DNA from specific chromosome regions is microdissection. First applied to polytene chromosomes of *Drosophila* salivary glands, this method has been extended to mammalian metaphase chromosomes, and with the development of PCR, it is a rapid and reasonably simple means of preparing a library of

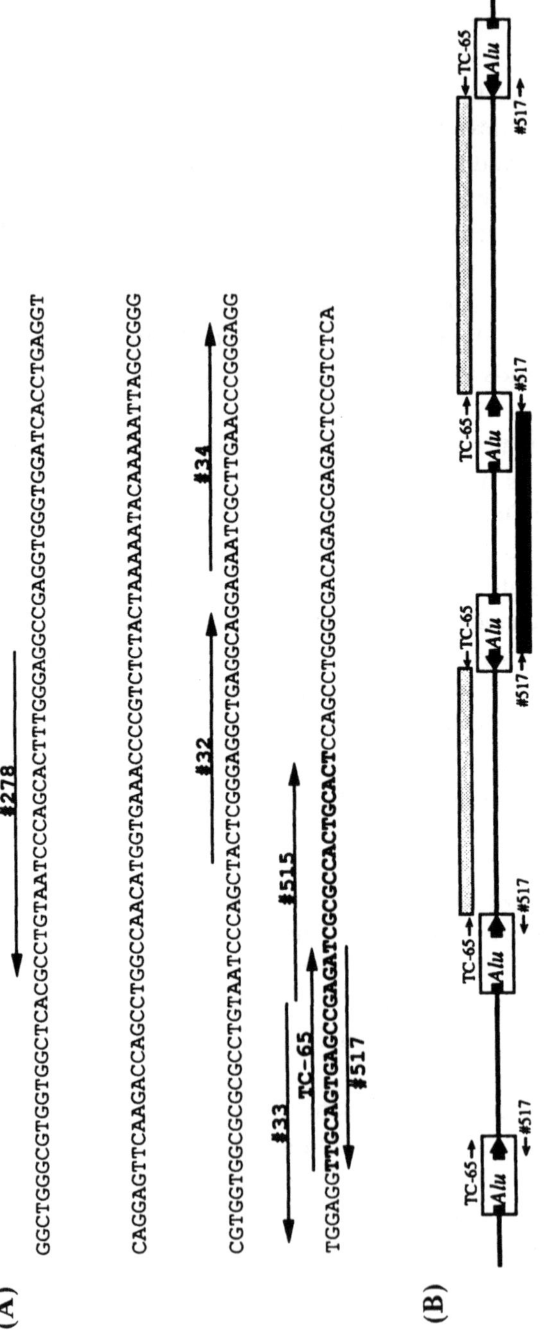
(A)
#278
GGCTGGGCGTGGTGGCTCACGCCTGTAATCCCAGCACTTTGGGAGGCCGAGGTGGGTGGATCACCTGAGGT
CAGGAGTTCAAGACCAGCCTGGCCAACATGGTGAAACCCCGTCTCTACTAAAAATACAAAAATTAGCCGGG
#32
#34
CGTGGTGGCGCGCCTGTAATCCCAGCTACTCGGGAGGCTGAGGCAGGAGAATCGCTTGAACCCGGGAGG
#33
TC-65
#515
#517
TGGAGGTTGCAGTGAGCCGAGATCGCGCCACTGCACTCCAGCCTGGGCGACAGAGCGAGACTCCGTCTCA
(B)
TC-65
Alu
#517

Figure 1

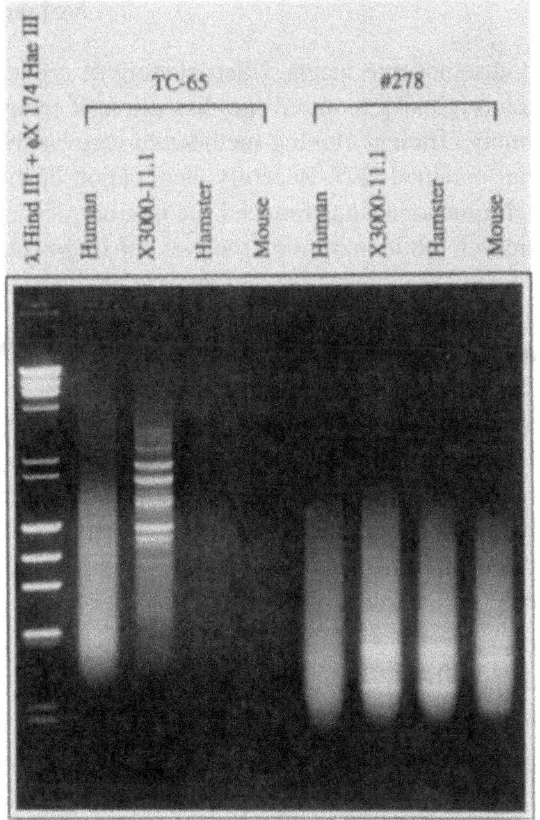

(C)

Figure 1 *Alu* polymerase chain reaction. Using oligonucleotide primers directed to sequences in the human *Alu* repetitive element, it is possible to amplify human DNA specifically from complex mixtures of nucleic acids (28). (A) The consensus *Alu* repeat from the human and the positions of seven primers tested. (B) The expected amplification products from the two human-specific primers, TC-65 and #517. Note that amplification with a single primer can only proceed from repeats in inverted orientation. Results from two primers (TC-65 and #278) are shown in C. Amplification with the TC-65 primer is human-specific, with no product observed in the hamster or mouse lanes, while #278 generates product in all DNA samples tested (human, mouse, and hamster). The lane marked X3000-11.1 is a hybrid cell line containing Xq24–qter as its only human material on a hamster cell background. TC-65 generates a complex pattern of products from this ~ 40-Mb region of the human genome. Thus, *Alu* PCR allows the isolation of fragments from specific genomic regions retained in hybrid cells.

fragments from specific chromosome bands. Microcloning of mouse and human chromosomes originally required the dissection of many chromosomes and extremely efficient cloning methods to recover the small amounts of material obtained (30). Recently, application of the PCR to microdissected chromosomes has reduced the number of dissected chromosomes required, while increasing the yield of fragments from a region. In this protocol, the microdissected material is digested with a restriction enzyme that cuts DNA frequently, giving small fragments. Then linker molecules are attached, either oligonucleotides (31) or cloning vectors (32) by ligation. Sequences in the added linkers are used to amplify the intervening material via PCR. With the addition of the laser microbeam for computer-aided microdissection, this method will allow the isolation of region-specific DNA fragments readily and rapidly (33). Again, such small clones can be used (individually or en masse) to isolate larger cloned fragments specific to the region from total human DNA libraries.

ISOLATION OF ADJACENT AND NEARBY DNA FRAGMENTS

The simplest method for isolation of DNA fragments adjacent to an original cloned marker is "chromosome walking." That is simply the use of the original probe to identify additional cloned DNAs that overlap it, followed by preparation of probes from the furthest end of the new clone and rescreening to continue the procession down the chromosome. This was first described by Bender et al. in *Drosophila* (34) and was usually carried out in cosmid libraries (average insert 35–40 kb); some quite impressive walks have been completed since (35). Two main difficulties are encountered with chromosome walks. The first is the tedium involved and the serial nature of the work. Each "step" in a walk required growing, mapping, and isolating a probe sequence from the prior clone (and often analyzing multiple clones to determine which extends furthest in the desired direction), followed by screening an entire library in order to begin the process again. Steps can take 1 week each and provide only a 10-or 20-kb extension. New cosmid and phage vectors incorporating bacteriophage T3 and T7 RNA promoters adjacent to the inserted DNA (pWE) (36) can reduce the cycle time for a single step by simplifying the preparation of probe. The other problem with chromosome walking involves the phenomenon of "unclonable" regions. Usually, unclonability is due to sequences found in the chromosome that are unacceptable to the *Escherichia coli* host, halting a walk in its tracks. Such sequences can be crossed by chromosome jumping (see below). Occasionally, sequences are encountered that contain too high a proportion of repetitive elements, and single-copy probes are difficult or impossible to obtain. This will also stop a walk.

Chromosome jumping was initially demonstrated by Francis Collins and Hans Lehrach in separate studies (37,38). The concept is to move down the chromosome in a more rapid manner than chromosome walking, by hopping to sequences a defined distance from the original clone. This is accomplished by the creation of a recombinant library of genomic DNA where inserts in the library are made up of DNA fragments that are separated in the genome by a large distance (see Fig. 2). Distances in such libraries are on the order of 100–200 kb. By probing the library with an initial probe, the clone identified will contain the original probe and the distant fragment. Jumping libraries are constructed by circularization (in the presence of a selectable marker such as sup F) of large DNA fragments that have been produced either by partial digestion with a frequently cutting restriction enzyme or by complete digestion with a rare-cutting enzyme (such as *Not* I), followed by redigestion with a second enzyme in such a way as to create clonable fragments. Inserts are separated by the selectable marker site initially used to create the large DNA fragments and circularization. In this way with a probe adjacent to a *Not* I site, it is possible to jump to the next *Not* I site that forms the opposite end of the *Not* I fragment in the genome. In a rare-cutter-jumping library, it is necessary to have a complementary linking library to facilitate crossing the rare-cut site at the opposite end. Linking libraries are prepared by isolating fragments containing rare-enzyme sites, typically by circularizing partially digested and size-selected DNA and then digesting with the rare cutter and cloning. Only fragments containing the site will be cloned, reducing the complexity considerably from that of a normal genomic library. The added advantage of *Not* I linking libraries is that the clones found in such libraries are greatly enriched in sequences from CpG islands, which are useful markers of expressed genes (see below). From the corresponding linking

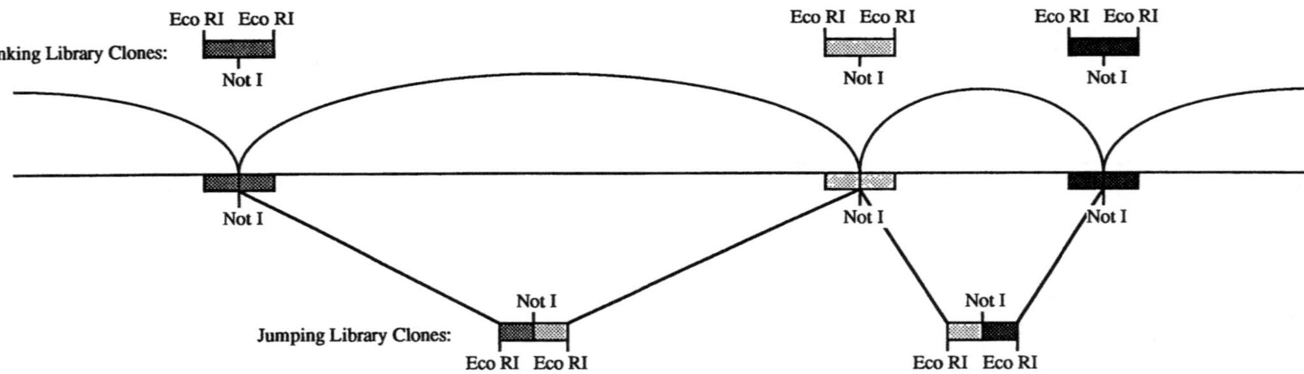

Figure 2 Chromosome jumping and linking with *Not* I. Jumping libraries can be constructed by circularization of long DNA fragments generated by *Not* I digestion (see text), providing cloned fragments with *Not* I sites internal to frequent cutters (such as *Eco* RI in this example). Linking libraries are made by selection of clones containing a rare cutting site (such as *Eco* RI fragments containing a *Not* I site in this example) and are complementary libraries, allowing the progression along a chromosome through the alternate screening of jumping and linking libraries.

clone, the next jump can be undertaken, as the restriction site is crossed. It was through the use of chromosome jumping that the journey down chromosome 7 from *met* to the CF gene was greatly accelerated.

More rapid means of walking down chromosomes are now possible through the use of libraries constructed in yeast artificial chromosome (YAC) vectors (39). The YAC system provides the ability to isolate large fragments of DNA (100–1000 kb) in yeast cells as stably replicated additional chromosomes. The vector provides selectable markers, a centromere for correct segregation of the chromosome at cell division, a replication origin, and telomeres to protect the ends of the chromosome. Libraries of human DNA and subregions from somatic cell hybrids have been developed in such vectors and represent extremely valuable resources (40,41). A single YAC insert of several hundred kilobases of human DNA provides the same ability to march down a chromosome as one or more chromosome jumps and has the benefit of retaining all the sequence in between, unlike jump clones. Furthermore, isolation of the equivalent contiguous DNA of a 200-kb YAC clone would require approximately 50 lambda clones of 20 kb each or 25 cosmid clones of 40 kb each and the subsequent mapping and ordering to determine overlaps. Obviously, chromosome walking carried out in the YAC clones offers similar considerable advantages. YACs represent a quantum leap in the isolation of linked DNA fragments.

Yeast artificial chromosome clones can be identified by DNA hybridization, as has been the standard procedure for clones in *E. coli* vectors. However, this has been found to be more difficult for clones in yeast. High-density colony screening is less sensitive in yeast owing to the smaller number of cells in a colony, the presence of fewer cloned molecules per cell and the difficulty in lysing yeast cells. Green and Olson (42) have devised a PCR-based screening protocol for a large-insert YAC library from human DNA. Screening through pools of YAC DNAs subjected to PCR with a set of primers directed to a single-copy sequenced region (as probe) allowed the identification of a final pool of 384 clones, which were then screened by hybridization to the PCR product. This method provides a rapid, simple alternative to screening all yeast colonies bearing YACs It also led to the concept of using PCR to characterize clones and regions. The sequence tagged site (STS) proposal (43) calls for identification of short regions of sequence spaced approximately 100 kb apart to form the backbone of the human genome map. Sequence at such sites can be used to direct the synthesis of PCR primers to identify YAC and other clones containing the region. Overlapping clones can be identified by the capacity to coamplify the same STS, while those STSs that are not shared define the end of the overlap. Also, the production of STSs from the ends of YAC clones offers the capability for walking in YACs through the PCR screening method.

Finally, a means for isolating DNA surrounding an initial probe through the isolation of large DNA fragments directly from pulsed-field gel electrophoresis has been developed. This entails isolation of the agarose in the vicinity of a hybridizing band from a pulsed-field gel, digestion, cloning, and clone identification. As multiple bands are encountered in a particular slice of such a gel, a means for identifying fragments from the specific band desired is necessary. This is provided by the use of DNA from a somatic cell hybrid with a reduced amount of the desired genome and screening the library constructed with repeat sequence probes specific for the DNA in question. This approach limits the number of similar-sized bands isolated from the desired genome. In the CF case, the Lehrach laboratory isolated new probes from a large pulsed-field gel fragment using a hybrid that retained a portion of human chromosome 7 in a rodent background (44). The use of PCR and species-specific repetitive primers (as in *Alu* PCR) may provide a more rapid means of achieving this goal. This method is of greatest interest when there is a reason to believe that the fragment to be isolated contains the disease gene.

HOW DO YOU KNOW WHEN YOU'RE THERE?

Although the isolation of DNA from the desired region represents a significant hurdle in the process of identifying disease genes, finding the gene from among the cloned sequences is equally challenging. In general, the primary test for candidate sequences involved in the disease phenotype is expression as an RNA transcript. The direct testing of DNA fragments for expression as RNA is difficult owing to the uncertainty of which tissue to test, the time of development to assay, and the possibility of a very low-abundance mRNA. Therefore, two other characteristics of DNA fragments have been applied to this problem. The first uses hybridization to DNAs derived from a variety of species (the "zoo blot") to detect regions of sequence conservation. This method was developed by Tony Monaco in the identification of the Duchenne muscular dystrophy (DMD) gene (45) and has proved to be quite useful in other searches, notably CF. The rationale involves the observation that coding sequences (exons) evolve at a much reduced rate in comparison to noncoding sequences (introns, DNA between transcription units, etc). Therefore, with reduced stringencies of hybridization and washing, it is likely that exon-containing probes will cross-hybridize between species. This has been the general observation for cloned genes, and it extends nicely to random DNAs. The method is significantly simpler than the use of Northern (RNA) hybridization. This is due to the use of DNA, where the abundance of all molecules is equal. A few widely disparate vertebrate species assayed by Southern hybridization can provide rapid, reasonably definitive results. It still represents a clone-by-clone method and is therefore labor intensive.

The second major method employs the observation that regions of mammalian genomes exhibiting high contents of the dinucleotide CPG are often found near promoter regions (46). The remainder of the genome is relatively deficient in CpG; therefore, a DNA fragment containing a high CpG content (generally assayed by the presence of rare-cutting restriction enzyme sites, which are rare due to the presence of CpG in their recognition sequence) is a good marker of a gene. Not all genes identified demonstrate such CpG islands, however, so while the presence of a rich CpG content is encouraging, its absence is not necessarily meaningful. Generally clones containing CpG islands are tested by zoo blot to confirm the presence of a nearby exon, as the islands are not usually found in exons.

Once convinced of cross-species hybridization, the DNA fragment is used to identify RNA transcripts by northern hybridization. Positive results here provide abundant information. The tissue and time of development that are positive allow the appropriate cDNA library to be assayed (or constructed) for the isolation of cDNA clones. Also, the tissue that is positive by this analysis allows speculation as to whether the fragment in question has the potential for representing the disease-causing gene. Expression in the appropriate tissue and lack of expression in some patients would represent two characteristics of the northern analysis that would provide evidence for the disease gene. cDNA clones isolated can be used to determine the sequence of the mRNA, allowing comparison to other cloned genes and speculation about the potential function of the protein encoded. Few examples exist, but the potential for studying a number of cDNAs before the actual one is found is fairly high. In the CF case, the CF gene was the fourth cDNA analyzed from the roughly 300-kb region covered by the walking and jumping clones. cDNAs can also be used to test patient samples for DNA rearrangements. This was of particular value in DMD, where a high proportion of patients exhibit deletions of portions of the gene (47). cDNA probes offer unique advantages in this regard, as they often cover large genomic regions (over 200 kb in CF, ~1.5 Mb in DMD). Using cDNA probes to isolate genomic clones allows a form of chromosome jumping between exons—this can be quite useful even if the cDNA is not the desired gene.

Mutations in the gene found on affected chromosomes are the hallmark of the genetic disease and must be consistently found with the phenotype. In DMD, with its high level of gene rearrangement, this was found readily. The CF case required the comparison of cDNA sequences from affected homozygous individuals to the normal sequence. In the CF case, the mRNA is expressed in the affected individuals, allowing cDNA to be utilized for this analysis. In other mutant genes, where no expression is found, sequence comparison requires the use of genomic DNA. With the capacity to rapidly and faithfully amplify and sequence DNA from affected individuals, this represents a much less difficult task than it was until quite recently (48). Additional methods to identify potential regions of mismatch between normal and mutant genes (49–51) provide the ability to direct sequence analysis to the relevant areas of the gene. Finally, the ultimate proof of the identity of a disease gene is the use of a normal copy to correct the defect. Where a cell culture system exists for such an assay, complementation of the defect is quite straightforward. The defect in Lesch-Nyhan syndrome, HPRT deficiency, for example, allows the rescue of HPRT-deficient cells with the normal cDNA (52). In nonhuman systems, the ultimate proof lies in the rescue of mutant individuals through the use of transgenic or gene-therapy methods (53). For humans, cell culture, the consistent finding of mutant genes in diseased individuals, and the probability that the candidate protein could cause the phenotype must suffice.

THE APPROACH OF THE HUMAN GENOME INITIATIVE

Each of the disease genes thus far identified required large, directed efforts focused on the individual gene and region. The major effort of these projects involved the isolation of DNA clones surrounding the region exhibiting linkage. While the ultimate aim of the human genome initiative is sequence of the entire genome, an intermediate, necessary goal is isolation of overlapping cloned DNAs covering each of the chromosomes. The availability of such banks of clones (within the next 5 years) will remove the difficulties involved in isolating DNA near disease genes and allow the emphasis to be placed on testing clones for their involvement in the disease process. Much of the methodology described here will be obsolete once sets of overlapping clones for each chromosome are collected and assembled. Although it is not explicit at this stage of the planning of the genome project, identification of transcripts from each region of each chromosome will also be an important intermediate goal. If achieved, this will allow only the transcription units to be analyzed in disease gene hunts, further reducing the effort. Finally, once the entire sequence is known, testing can potentially proceed electronically, searching coding sequences in the relevant region for possible homology to genes thought to be similar in function to the defective one.

The genome project has already demonstrated its capacity to drive the technology necessary for disease gene identification through its support. The development of large insert cloning vectors (YAC and others), the support of large-scale mapping methods, novel sequencing technologies, and clone collection and distribution will radically alter the methods of disease gene isolation. Perhaps the only remnant of the current methods utilized for genetic disease analysis that will be found in 10 or 15 years will be the collection of pedigrees and ascertainment of phenotype.

ACKNOWLEDGMENTS

I wish to acknowledge the support of Dr. C. Thomas Caskey, in whose group this work was done, as well as my many colleagues and collaborators for their instruction and advice. This work was supported by the Howard Hughes Medical Institute and a grant from the US Department of Energy, DE-FG05-88ER60692.

REFERENCES

1a. Rommens JM et al. Identification of the cystic fibrosis gene: chromosome walking and jumping. Science 1989; 245:1059-1065.
1b. Riodan JR, et al. Identification of the cystic fibrosis gene: cloning and characterization of complementary DNA. Science 1989; 245:1066-1073.
1c. Kerem B-S et al. Identification of the cystic fibrosis gene: genetic analysis. Science 1989; 245:1073-1080.
2. Orkin SH. Reverse genetics and human disease. Cell 1986; 47:845-850.
3. Monaco AP, et al. Isolation of candidate cDNAs for portions fo the Duchenne muscular dystrophy gene. Nature (1986) 323:646-650.
4. Royer-Polara B, et al. Cloning the gene for an inherited human disorder—chronic granulomatous disease—on the basis of its chromosomal location. Nature (1986) 322:32-38.
5. Friend SH, et al. A human DNA segment with properties of the gene that predisposes to retinoblastoma and osteosarcoma. Nature (1986) 323:643-646.
6. Call K, et al. Isolation and characterization of a zinc finger polypeptide gene at the human chromosome 11 Wilm's tumor locus. Cell (1990) 60:509-520.
7. Gusella JF, et al. A polymorphic DNA marker genetically linked to Huntington's disease. Nature (1983) 306:234-238.
8. Reeders ST, et al. A highly polymorphic DNA marker linked to adult polycystic kidney disease on chromosome 16. Nature (1985) 317:542-544.
9. Bodmer WF, et al. Localization of the gene for familial adenomatous polyposis on chromosome 5. Nature (1987) 328:614-616.
10. Davies KE, et al. Linkage analysis of myotonic dystrophy and sequences on chromosome 19 using a cloned complement 3 gene probe. J Med Genet (1983) 20:259-263.

11. Sherrington R, et al. Localization of a susceptibility locus for schizophenia on chromosome 5. Nature (1988) 336:164–167.
12. Barker D, et al. Gene for von Recklinghausen neurofibromatosis is in the pericentromeric region of chromosome 17. Science (1987) 236:1100–1102.
13. Gusella JF, et al. 1983; op cit.
14. White RL, et al. A closely linked genetic marker for cystic fibrosis. Nature 1985; 318:382–384.
15. Gusella JF. Locaiton cloning strategy for characterizing genetic defects in Huntington's disease and Alzheimer's disease. FASEB J 1989; 3:2036–2041.
16. Bates GP, et al. Isolation of a telomeric clone from chromosome 4p likely to contain the mutant form of the Huntington;s disease gene. Am J Hum Genet (1989) 45:A173.
17. Chakravarti A, Li CC, Buetow KH. Estimation of the marker gene frequency and linkage disequilibrium from conditional marker data. Am J Hum Genet (1984) 36:177–186.
18. Donis-Keller H, et al. A genetic linkage map of the human genome. Cell (1987) 51:319–337.
19. Poustka AM, Lehrach H, Williamson R, Bates G. A long-range restriction map encompassing the cystic fibrosis locus and its closely linked genetic markers. Genomics (1988) 2:337–345.
20. Gusella JF, et al. Isolation and localization DNA segments from specific human chromosomes. Proc Natl Acad Sci USA (1980) 77:2829–2833.
21. Shows TB, Naylor SL, Sakaguchi AY. Mapping the human genome, cloned genes, DNA polymorphisms and inherited disease. In: Hirschorn K, Harris H. eds. Advances in human genetics, Vol 12. New York:Plenum Press, 1982:341–452.
22. Southern EM. Detection of specific sequences among DNA fragments separated by gel electrophoresis. J Mol Biol (1975) 98:503–517.
23. Shih C, Weinberg RA. Isolation of a transforming sequence from a human bladder carcinoma cell line. Cell (1982) 29:161–169.
24. Britten RJ, Baron WF, Stout DB, Davidson EH. Sources and evolution of human *Alu* repeated sequences. Proc Natl Acad Sci USA (1988) 85:4770–4774.
25. Goss SJ, Harris H. New method for gene mapping in human chromosomes. Nature (1975) 255:680–682.
26. Housman DE, Nelson DL. Use of metaphase chromosome transfer for Mammalian gene mapping. In: Kucherlapati R. ed. Gene transfer. New York: Plenum Press, 1986:95–115.

27. Saiki RK, et al. Primer-directed enzymatic amplification of DNA with a thermostable DNA polymerase. Science (1988) 239:487- 489.
28. Nelson DL, et al. *Alu* polymerase chain reaction: a method for rapid isolation of human-specific sequences from complex DNA sources. Proc Natl Acad Sci USA (1989) 86:6686-6690.
29. Gray JW, et al. High-speed chromosome sorting. Science (1987) 238: 323-329.
30. Bates GP, Wainwright BJ, Williamson R, Brown SD. Microdissection of and microcloning from the short arm of human chromosome 2. Mol Cell Biol (1986) 6:3826-3830.
31. Johnson DH. Molecular cloning of DNA from specific chromosomal regions by microdissection and sequence-independent amplification of DNA. Genomics (1990) 6:243-251.
32. Lüdecke H-J, Senger G, Claussen U, Horsthemke B. Cloning defined regions fo the human genome by microdissection of banded chromosomes and enzymatic amplification. Nature (1989) 338:348-350.
33. Monajembashi S, et al. Microdissection of human chromosomes by a laser microbeam. Exp Cell Res (1986) 167:262-265.
34. Bender W, Spier P, Hogness D. Chromosomal walking and jumping to isolate DNA from the ace and rosy loci and the bithorax complex in *Drosophila melanogaster*. J Mol Biol (1983) 168:17-33.
35. Steinmetz M, Stephan D, Fischer-Lindahl K. Gene organization and recombinational hotspots in the murine major histocompatibility complex. Cell (1986) 44:895-904.
36. Wahl GM, et al. Cosmid vectors for rapid genomic walking, restriction mapping and gene transfer. Proc Natl Acad Sci USA (1987) 84: 2160-2164.
37. Collins FS, Weissman SM. Directional cloning of DNA fragments at a large distance from an initial probe: a circularization method. Proc Natl Acad Sci USA (1984) 81:6812-6816.
38. Poustka A, et al. Construction and use of human chromosome jumping libraries from *Not* I-digested DNA. Nature (1987) 325:353-355.
39. Burke DT, Carle GF, Olson MV. Cloning of large segments of exogenous DNA into yeast by means of artificial chromosome vectors. Science (1987) 236:806-812.
40. Brownstein BH, et al. Isolation of singly-copy human genes from a library of yeast artificial chromosome clones. Science (1989) 244:1348-1351.

41. Little RD, et al. Yeast artificial chromosomes with 200–to 800–kilobase inserts of human DNA containing HLA, V kappa, 5S, and Xq24–Xq28 sequences. Proc Natl Acad Sci USA. (1989) 86:1598–1602.
42. Green E, Olson MV. Systematic screening of yeast artificial-chromosome libraries by use of the polymerase chain reaction. Proc Natl Acad Sci USA (1990) 87:1213–1217.
43. Olson M, Hood L, Cantor C, Botstein D. A common language for physical mapping of the human genome. Science (1989) 245:1434– 1435.
44. Michiels F, Burmeister M, Lehrach H. Derivation of clones close to met by preparative field inversion gel electrophoresis. Science (1987) 236:1305–1308.
45. Monaco AP, et al. Nature 1986; op. cit.
46. Bird A. CpG islands as gene markers in the vertegrate nucleus. Trends Genet (1987) 3:342–347.
47. Baumback LL, et al. Molecular and clinical correlations of deletions leading to Duchenne and Becker muscular dystrophies. Neurology (1989) 39:465–474.
48. Gibbs RA, et al. Identification of mutations leading to the Lesch-Nyhan syndrome by automated direct DNA sequencing of intro amplified cDNA. Proc Natl Acad Sci USA (1989) 86:1919–1923.
49. Gibbs RA, Caskey CT. Identification and localization of mutations at the Lesch-Nyhan locus by ribonuclease A cleavage. Science (1987) 236:303–305.
50. Grompe M, Munzny DM, Caskey DT. Scanning detection of mutations in human ornithine transcarbamoylase by chemical mismatch cleavage. Proc Natl Acad Sci USA (1989) 86:5888–5892.
51. Sheffield VC, Cox DR, Lerman LS, Myers RM. Attachment of a 40-base-pair G + C-rich sequence (GC-clamp) to genomic DNA fragments by the polymerase chain reaction results in improved detection of single-base changes. Proc Natl Acad Sci USA (1989) 86:232–236.
52. Brennand J, Konechi DS, Caskey CT. Expression of human and Chinese hamster hypoxanthine-guanine phosphoribosyltransferase cDNA recombinants in cultured Lesch-Nyhan and Chinese hamster fibroblasts. J Biol Chem (1983) 258:9593–9596.
53. Readhead C, et al. Expression of a myelin basic protein gene in transgenic shiverer mice: correction of the dysmyelinating phenotype. Cell (1987) 48:703–712.

TECHNIQUES FOR GENE MAPPING

5

Hypervariable Minisatellites and Their Use in Animal Breeding

Michel Georges

Genmark Inc.
Salt Lake City, Utah

INTRODUCTION

The possibilities offered by the so-called "reverse genetics" approach (Orkin, 1986), which were anticipated nearly 10 years ago by Botstein et al. (1980), have since been fully proven by the discovery of genetic markers for a variety of human diseases, followed by the cloning of some of the involved genes. A nearly complete human genetic map, fruit of a worldwide collaborative effort, has been the basis for this success. Besides monogenic traits, the first to be tackled, polygenic entities are now being studied. The mapping of quantitative trait loci (QTL) in plant species, for which excellent genetic maps are available, testifies to the effectiveness of the strategy to unravel the genetic determinism of complex traits (Paterson et al. 1988; Martin et al., 1989).

The potentialities of the reverse genetic approach in animal breeding have been considered very early from a theoretical point of view. Mapping genes involved in economically relevant mono-or polygenic entities would offer the possibility for marker-assisted selection (MAS), expected to improve genetic response by affecting both accuracy and time of selection (Soller and Beckmann, 1982; Smith and Simpson, 1986).

Introgression of genes from one stock into another has been considered as a specific application of MAS (Beckmann and Soller, 1987). Moreover, mapping of the relevant genes would be the first step toward their isolation, the analysis of their mode of action, and their manipulation by the continuously improving mutagenesis and gene transfer methods.

Putting these ideas into practice, however, has been hampered by the shortage of genetic markers for our main domestic species. In cattle, for instance, until a few years ago, bovine markers were limited to 10 blood group systems, serologically defined BoLA antigens (major histocompatibility complex), and a few dozen biochemical markers. Recently, Fries et al. (1989), in their review on the bovine gene map, reported DNA sequence polymorphisms (DSP) at 23 bovine loci. Even benefitting from the wealth of comparative data from the human and rodent sphere, building complete animal genetic maps based on classical restriction fragment length polymorphisms (RFLP) could be regarded a Herculean task for the limited scientific community involved. The observation of a reduced genetic variation, measured in terms of nucleotide diversity, and resulting probably from the population structure imposed by breeding strategies, could only make the task more difficult (Georges et al., 1987a; Hilbert et al., 1989).

In 1985, Alec J. Jeffreys, from Leicester University, identified a new class of human DNA sequences: the families of hypervariable minisatellites. In the human, these sequences proved to be an invaluable source of highly informative genetic markers with a broad range of applications in fundamental research, such as mapping studies and cancer genetics, as well as in applied sciences, including paternity diagnosis and forensic medicine. This discovery raised the hope among animal geneticists that, if the same sort of sequences could be identified in the genomes of our domestic species, a large number of markers might be generated in a reasonable time span, providing a new and powerful breeding tool: genetic maps of domestic animals.

The purpose of this chapter is to summarize the knowledge that has been gained in the last 4 years about families of hypervariable minisatellites in humans and domestic animals and to evaluate whether these sequences may fulfill the expectations they raised among animal breeders.

A NEW COMPONENT OF THE HUMAN GENOME: HYPERVARIABLE MINISATELLITES

Families of Hypervariable Minisatellites (Jeffreys et al., 1985)

Studying an intron of the human myoglobin gene, Jeffreys identified a 33-bp sequence repeated four times in a head-to-tail or "tandem" arrangement. This so-called "minisatellite" showed similarity with sequences previously identified within the human α-globin gene cluster (Goodbourn et al., 1983), 5′ to the human insulin gene (Bell et al., 1982) and 3′ to the human Ha-Ras protooncogene (Capon et al., 1983), characterized by a substantial genetic polymorphism due to variation in the number of tandem repetitions. Looking for the presence of other related and potentially polymorphic minisatellites in the human genome, Jeffreys used the 33-bp myoglobin repeat in low-stringency hybridization experiments and demonstrated that this probe cross-hybridized indeed with several related sequences.

Cloning and sequencing of some of these cross-reacting sequences demonstrated that they were characterized by the presence of related minisatellites with repeat units varying in length between 16 and 64 bp, but sharing with the myoglobin probe a consensus, G≡C-rich "core" sequence: GGAGGTGGGCAGGAXG, defining the first identified human family of hypervariable minisatellites.

Four of the eight initially described minisatellite loci were showing a substantial degree of genetic polymorphism (heterozygosities between 0.47 and O.73), due to allelic variations in the number of repeats. This exceptional polymorphism is thought to result from the very high mutation rate characterizing minisatellite sequences. The mutation rate a some highly variable minisatellite loci has been directly measured in human pedigrees and shown to reach 5% per gamete for the most unstable loci. Overall, mutation rate and observed variability at different loci appear to correlate in accord with the neutral mutation-random drift hypothesis. Neutral mutation rates therefore appear to be sufficiently high to maintain the level of variability seen at these hypervariable loci (Jeffreys et al., 1988). Absence of significant within-family clustering of neomutations, never shown to be shared between sibs, and comparable mutation rates in sperm and oocytes suggested that the mutational event probably occurs at the latest stages of gametogenesis (Jeffreys et al., 1988). The most prominent model is unequal crossing-over between homologous chromosomes at meiosis. The demonstration of absence of recombination between markers flanking a neomutated allele indicates, however, that other mechanisms might also be involved (Wolff et al., 1988,1989). Moreover, the demonstration of somatic mutations in lymphoblastoid cell lines and tumors indicates that mutations at minisatellite loci can occur in the absence of meiosis (Armour et al., 1989).

The function of minisatellite sequences, if any, is essentially unknown. Although some coding sequences exhibit a minisatellite-type structure, sometimes even showing genetic polymorphism (Swallow et al., 1987), the majority of identified minisatellites are probably noncoding. The homology observed between the "core" sequences defining Jeffreys' as well as other families of hypervariable minisatellites and the *Escherichia coli* "chi" sequence known to mediate recBC-dependent recombination favored the hypothesis that those sequences might promote recombinational events also in eukaryotes. Based on the neutral mutation–random drift hypothesis, the mean mutation rate has been estimated at $\pm$ 0.5–1.5 $\times$ 10^{-4} per gamete for a $\pm$ 1 kb-long minisatellite. This value has to be compared with a rate of homologous recombination of 10^{-5} per kb (1 cM $\approx$ 10^{6} bp). If unequal crossing-over between homologous chromosomes at meiosis is indeed the most important mutational mechanism, hypervariable minisatellites could be hotspots for meiotic recombination (Jeffreys et al., 1985). However, no direct evidence is available at the present time to confirm this hypothesis. In one instance, a minisatellite has been shown to be closely associated with a sequence exhibiting enhancer activity (Cohen et al., 1987). Moreover, Sen and Gilbert (1988) have shown that guanine-rich motifs resembling minisatellites form four-stranded structures with parallel running strands. They speculated that this self-recognition serves to bring the four homologous chromatids together during meiosis.

Multilocus DNA Fingerprints

When using minisatellites as probes in Southern blot hybridization experiments at low stringency, very complex, individual specific patterns, known as multilocus DNA fingerprints, are obtained. With Jeffreys 33.15 and 33.6 probes, for instance, DNA fingerprints are characterized by, respectively, 15 and 14 resolvable bands above 4-kb size limit, under which the patterns are too complex to be properly interpreted. The probability to be heterozygous for the corresponding loci increases from 86% to 96% with increasing fragment size (Jeffreys and Wilson, 1985).

Later, several other probes were identified revealing similar multilocus DNA fingerprints in the human. These probes include the hypervariable region 3′ to the human α-globin gene (Jarman et al., 1986), a minisatellite characterizing the protein III gene of bacteriophage M13 (Vassart et al., 1987), a human Y-specific satellite III sequence (Fowler et al., 1987), a mouse probe related to part of the *Drosophila Per* gene (Georges et al., 1987b), as well as several other human minisatellites. The Vassart study involving the M13 minisatellite had the merit, besides demonstrating a surprising sequence similarity between a minisatellite in the phage protein III gene and mammalian hypervariable minisatellites, that avoiding salmon sperm DNA in hybridization conditions substantially increased the complexity and quality of the DNA fingerprints. This was attributed to the richness of this competitor DNA in minisatellite-type sequences, hence masking the fingerprint pattern when used in the hybridization cocktail.

It is reasonable to assume that those different probes are each recognizing a different family of hypervariable minisatellites. Superimposition of fingerprints obtained with a single blot hybridized successively to different probes will usually show no significant overlap between patterns. The repeated motifs characterizing these multilocus DNA fingerprinting probes have a high G≡C content in common. Another type of probes based on simple quadruplet repeats (sqr), originally isolated from female-specific snake satellite DNA, from the type GAPyA, as well as simple triplet and doublet repeats such as CAC, TCC, and CT, were also shown to reveal highly informative DNA finger-print like patterns (Ali et al., 1986; Schafer et al., 1988).

The interindividual variation of these patterns is such that the probability of matching fingerprints for two unrelated individuals is, for instance, $\ll 5 \times 10^{-19}$ when combining probes 33.15 and 33.6. For full-sibs this probability is $\pm\ 10^{-8}$ (Jeffreys and Wilson, 1985). The term “individual-specific" DNA fingerprints is, as a consequence, and with the exception of monozygotic twins, fully justified, making these probes outstanding tools for forensic medicine. Another obvious application is paternity diagnosis. Again, combining probes 33.15 and 33.6, the probabilities to miss an incorrect paternity are $\approx 4 \times 6^{-8}$ and $\approx 8 \times 10^{-4}$, if real and putative father are unrelated or sibs, respectively (Jeffreys and Wilson, 1985). The fairly high mutation rates characterizing the explored sequences, without being a major problem relative to their use in forensics and paternity diagnosis, must, however, be considered in the interpretation or DNA fingerprints. Besides these major applications, the individual specific character of the multilocus DNA fingerprints has led to other, more fundamental applications, such as determination of twin zygosity (Hill and Jeffreys, 1985), identification of posttransplant populations (Thein et al., 1986), and demonstration of spontaneous XX/XY chimerism (Farber et al., 1989).

The possibility of using the DNA fingerprints for a quick, cheap genome scanning in linkage studies was very attractive. Using the probes 33.15 and 33.6, Jeffreys could study the segregation of two apparently independent sets of human minisatellites adding up to 34 unlinked loci in the offspring of a given individual. The minimal locus-to-locus spacing was estimated at 30 cM, favoring the hypothesis of random dispersal of minisatellites across the human genome (Jeffreys et al., 1986). The results of a linkage study using recombinant inbred mice also favored the hypothesis of random distribution of minisatellites, without preferential association with centromeres or telomeres, at least in mice (Jeffreys et al., 1987). Note, however, that to the best of our knowledge, no linkage relationship has yet been demonstrated in the human using this approach.

Locus-Specific Variable Number of Tandem Repeat Markers

Although it is potentially useful to be able to explore a large number of hypervariable loci simultaneously, for some applications the multilocus DNA fingerprints have major drawbacks. In forensic medicine and paternity diagnosis, for instance, they may not be the ultimate tool. Because of the low stringency at which one has to work, DNA fingerprints are not always perfectly reproducible. Some explored systems may be at the borderline of cross-hybridization in the used conditions and will sometimes light up, other times not. The only way to reliably confront DNA samples is to run them altogether in a unique experiment. This means that the establishment of data bases is made very complicated when using multilocus DNA fingerprints.

In studies aimed to map genetic defects, codominantly transmitted diseases may be the only ones tractable, because it is difficult to combine results obtained on different pedigrees. Indeed, only for these entities might a single human family contain enough information to conclude for or against linkage. Moreover, if no linkage is found, and even if a significant proportion of the genome may have been explored using this approach, we cannot "exclude" the target gene from any part of the genome and must start from zero when using available locus-specific markers.

Most of these drawbacks could be circumvented if using the minisatellite probes at high stringency, exploring only one marker system at a time: the so-called variable number of tandem repeat (VNTR) markers. To isolate such VNTR markers, several laboratories have been trying to clone single members of the families of hypervariable minisatellites. Nakamura et al. (1987), in R. White's team have been particularly successful along those lines. Using a collection of previously identified minisatellites to screen cosmid libraries, they were able to isolate several hundred human VNTR type markers. Approximately 20% of the selected cosmids did show VNTR-type polymorphism, with heterozygosities often exceeding 70%. While Nakamura and Jeffreys in his original study were screening whole genomic libraries in cosmid and lambda vectors, respectively, another approach was based on the direct cloning of the high-molecular-weight bands of multilocus DNA fingerprints (Wong et al., 1986, 1987). The rationale behind this strategy, besides allowing targeting of a specific fingerprint band, is the fact that with the four-cutters used to generate fingerprints, the proportion of restriction fragments above a size limit of, for instance, 2kb is very low. Following Bishop et al. (1983), and assuming an exponential distribution of restriction fragment lengths, the fragments ≥ 2 kb represent about 10^{-3} of the total number of MboI fragments or $\pm 10^4$. In consequence, a library made up of this fragments is enriched in minisatellites relative to a total genomic library generated with the same enzyme. Moreover, Jeffreys et al. (1985) suggested that the mutation rate of minisatellites may increase with minisatellite length. According to this hypothesis, minisatellite loci characterized by alleles with a large number of tandem repeats and corresponding to the resolvable part of the DNA fingerprints may be the most polymorphic ones. Although only a limited number of VNTR markers have been isolated using this strategy, all of them were indeed showing heterozygosities ≥ 90%.

Moreover, several VNTR type markers were identified serendipitously when studying a number of genes such as the insulin gene (Bell et al., 1982), the Ha-Ras gene (Capon et al., 1983), the globin genes (Goodbourn et al., 1983; Jarman et al., 1986; Jarman and Higgs, 1988), the PUM gene (Swallow et al., 1987), the retinoblastoma gene (Wiggs et al., 1988), the human Factor VII gene (O'Hara and Grant, 1988), the pseudoautosomal region of the human sex chromosomes (Simmler et al., 1987), the apolipoprotein B gene (Knott et al., 1986), the collagen type II gene (Stoker et al., 1985), and others. Except for the last two, having A=T-rich repeat units, all the others were from the G≡C-rich type. It is noteworthy that one of the first identified human DNA polymorphism (using an anonymous probe) was a VNTR marker (Wyman and White, 1980).

As expected, VNTR markers have proven very useful in a wide variety of situations. In paternity diagnosis and forensic medicine, they are now preferred to the multilocus DNA fingerprints. The main impact of VNTR markers, however, has been in mapping studies. The hundreds of VNTR markers now available contribute significantly to the coverage of the human genome. They have contributed to the mapping of several disease genes. Besides classical linkage studies, search for loss of heterozygosity at VNTR loci has proven particularly efficient to map potential antioncogenes.

However, analysis of the available genetic maps of human chromosomes clearly indicates that the distribution of VNTR markers is not as random as initially deduced from linkage studies performed with human multilocus DNA fingerprints. VNTRs show an obvious tendency to be clustered toward telomeric regions (Lathrop et al., 1988a,b; Nakamura et al., 1988a–c; O'Connell et al., 1987, 1989). This has been confirmed by in situ hybridization experiments showing that at least four or the six mapped minisatellites were located at terminal G-bands of human autosomes. The two others were located toward the ends of chromosome 1, although not terminal. It is noteworthy, as pointed out by Royle et al. (1988), that there is evidence for linkage map expansion toward the ends of chromosomes, which might be related to the high density of minisatellites in these regions.

There is thus an apparent contradiction between the distribution of minisatellites in human and mouse, where at least 8 of 13 VNTRs detected were shown to have an interstitial map location (Jeffreys et al., 1987).

Although these facts do not affect the quality of VNTRs as markers when considered individually, it is more than questionable whether a complete map based on them will ever be available, and this urges for the search for other highly informative human markers.

HYPERVARIABLE MINISATELLITES IN DOMESTIC ANIMALS

Cross-Species Conversation of Families of Hypervariable Minisatellites Makes DNA Fingerprinting in Animals Easy

As previously mentioned, animal geneticists realized that if families of hypervariable minisatellites could be recognized in animals, this might provide them with an invaluable tool having a broad range of applications: individual identification, paternity diagnosis, behavioral studies, and eventually a means to improve selection procedures through the reverse genetic approach.

Fortunately, the probes known to reveal DNA fingerprints in the human were shown to recognize similar sequences in the genomes of the majority of animals tested (Wetton et al., 1987; Burke and Bruford, 1987; Jeffreys and Morton, 1987; Vassart et al., 1987 Georges et al., 1988a; Gyllensten et al., 1989). Some probes were even shown to work in plants (Ryskov et al., 1988; Dallas, 1988; Rogstad et al., 1988). The remarkable conservation of families of hypervariable minisatellites raises the question of their relatedness throughout evolution. Remember that a minisatellite in the bacteriophage M13 genome shows similarities with animal and plant minisatellites. Are these "similar" sequences homologs of one another that can be traced back to a common ancestor or does concerted evolution also play a role? Table 1 lists the probes known by the author to work in some of our domestic species.

The pUCJ probe is composed of tandem repetitions of Jeffreys 16-bp "core sequence," the 15th base being adenosine (Georges et al., 1988a). While Jeffreys' 33.15 probe is hard to use in cattle probably because of a cross-reaction with bovine satellite 1.720, the pUCJ sequence, which has only 8 out of 16 pairing bases with this satellite compared to 11, gives useful patterns when the stringency is increased sufficiently.

The animal patterns obtained are essentially similar to the ones found in the human: highly complex, individual, specific DNA fingerprints characterized by a large number of cross-hybridizing bands of which only part, above a certain size limit, are resolved well enough to be usable. The complexity of the patterns, however, seems to be a function of species as well as species-probe combination. All proves used on chicken DNA, for instance, yielded very complex and informative DNA fingerprints, while the bovine DNA fingerprints were almost always characterized by a lower number of fragments in the resolvable region and are therefore relatively less informative.

The genetic polymorphism characterizing these DNA fingerprints, measured, for instance, by the mean probability x that a fragment present in one individual is also present in another, randomly chosen individual, has usually been shown slightly lower compared to human fingerprints. Jeffreys reports x values increasing from 0.27 to 0.08 with increasing fragment size in the human (Jeffreys and Wilson, 1985). He found mean values of 0.46 in dogs and cats using the same 33.15 and 33.6 probes (Jeffreys and Morton, 1987), while we found values increasing from 0.27 to 0.76 when performing a study including four probes and several domestic species (Georges et al., 1988a). At least part of this reduced variation can probably be attributed to higher inbreeding resulting from population structure imposed by breeding strategies. Note that a similar, lower genetic variation was found in cattle when determining nucleotide diversities from site polymorphism detected as RFLPs. Belgian blue cattle were estimated to be heterozygous for 1/1450 nucleotides, a value significantly lower than the 1/500 value found in humans. In this study, the mutation rate at the cytosines of CG

Table 1 Minisatellite Probes Known to Reveal Multilocus DNA Fingerprints in Domestic Species

	33.15	33.6	M13	pUCJ	3'HVR	pINS	EFD134.7	pSP2.5EI	Rb	pS3
Cow	–		+	+	+	+	+	+	+	
Horse			+	+	–			+		+
Pig			+	+	–			+		+
Dog	+	+	+	+	–			+		
Cat	+	+								
Chick	+	+	+	+	+			+		
Fish			+	+	+			+		

References for probes 33.15 and 33.6 ae Jeffreys et al., 1985; for probes M13, pUCJ, 3'HVR, pINS, EFD134.7, Georges et al., 1988a and in press; for probe pSP2.5EI, Georges et al., 1988a, and Perret et al., in press; for pS3, Coppieters et al., in press; Rb (human retinoblastoma) is unpublished.

dinucleotides was estimated about 10 times higher than that of other nucleotides (Georges et al., 1987a; Hilbert et al., 1989).

Methods to estimate the breeding value of an animal use information from relatives. As a matter of fact, keeping track of familial relationships has always been one of the major concerns of animal breeders, and parentage control is now widely used for several domestic species. DNA fingerprinting has potential use in this regard. Combining results from three different probes, the probabilities of missing a wrong paternity were estimated in cattle, dog, and horses, at $\pm 3 \times 10^{-4}$ when real and putative father are unrelated and at $\pm 3 \times 10^{-2}$ when real and putative father are full-sibs (Georges et al., 1988a). The latter values approach those obtained by more classical means, such as blood group systems and biochemical polymorphisms, in case of unrelated real and putative parents, demonstrating the power of the method. The feasibility of DNA fingerprinting for parentage control in animals has been demonstrated by solving paternity disputes in dogs (Morton et al., 1987; Georges et al., 1988b) as well as other species (Georges et al., unpublished). However, and as in the human, multilocus DNA fingerprints suffer from a nonperfect reproductibility which will probably hamper their widespread use in animal parentage control. Kuhnlein et al. (1989) have used DNA fingerprint obtained with M13 as a tool for determining genetic distances between strains of poultry.

DNA Fingerprints to Map Economic Trait Loci

Since the studies of Bakewell, at the end of the eighteenth century, animal breeders have manipulated genes by artificial selection. Despite the development of sophisticated biometrical methods, quantitative geneticists have been dealing with a "black box," essentially without knowledge of the molecular basis upon which they were acting. Identification of the underlying genes, or economic trait loci (ETL), is expected to further improve selection response by affecting both accuracy and time of selection (Soller and Beckmann, 1982; Smith and Simpson, 1986). Moreover, it will increase our understanding of the mechanisms of artificial selection and eventually allow us to manipulate the involved genes, once isolated, by mutagenesis and gene transfer techniques.

One way to identify those ETL is through "reverse genetics." In a first step, this entails mapping of the ETL by identification of linked genetic markers. To perform an efficient linkage analysis, several hundred markers covering as much as possible of the genome are required. Unfortunately, and at the present time, few polymorphic markers are available for our domestic species. Therefore, the possibility of using DNA fingerprinting to explore a large number of loci simultaneously was particularly attractive to animal geneticists, especially because the main limit of this approach in the human, i.e., the difficulty of combining data gathered on different pedigrees, does not apply in domestic species. Indeed, very large animal pedigrees are available or can even be designed at will. Theoretically, thus, recessive as well as complex traits should be amenable to study following this strategy.

Along these lines, and in order to test this possibility, we performed a linkage study in cattle (Georges et al., 1990) using five DNA fingerprint probes (M13, pUCJ, 3'HVR, pINS, EFD134.7) in five paternal half-sib groups for a total of 540 individuals. Two of the pedigrees were segregating for the "muscular hypertrophy (MH)" locus; one was segregating at the "roan" locus. The MH gene is known to generate the double-muscled phenotype in those breeds (such as the Belgian blue cattle breed) harboring the partially recessive "mh" allele. Homozygous mh/mh animals are "double muscled," showing a 20% increase of muscle mass, all other organs being reduced in size. mh/+ and +/+ animals are "conventional" (Hanset and Michaux, 1985a,b). The "roan" locus is a coat color-determining gene characterized by two alleles with intermediate inheritance; black (r^+) and white (R); the heterozygous animals are qualified as "blue" (Hanset, 1985). Our interest in this locus stems from its potential role in "white heifer disease" (Hanset, 1965).

Analysis of the obtained DNA fingerprints showed that in our conditions the segregation of a mean of 7.6 clearly resolvable bands could be traced per individual and per probe. This figure may appear low compared to other studies. This may be attributed partly to the relatively lower complexity observed for bovine DNA fingerprints compared to other species, but also to the fact that we were very careful before considering a fingerprint band appropriate for linkage studies.

The fingerprint bands were shown to segregate as simple, highly informative mendelian entities. Indeed, a mean of 73% of the offspring was informative for segregation of the fingerprint bands. Two neomutated bands were found in a total of ± 9000 offspring bands studied, giving a mean mutation rate of 2.2×10^{-4} per gamete, in reasonable agreement with Jeffreys' report of ± $0.5–1.5 \times 10^{-4}$ in humans. Moreover, we observed one instance of germline mosaicism, apparently due to a mutation burst in the germline generating a collection of neomutated alleles while completely eliminating the original allele still present in somatic tissue.

Besides allelic pairs (reducing the number of explored loci to 6.3 per probe), linkage analysis revealed a substantial number of closely linked fingerprint bands both "within" (reducing the number of independently segregating loci further to 5.7 per probe) and "between" probes. Moreover, the distribution of recombination rates between all possible pairs of minisatellites (excluding allelic pairs), without considering lod score values, showed a peak at low recombination rates ($\Theta \leq 0.05$). These observations favor a nonrandom distribution of minisatellites across the bovine genome. From the observed distribution of linked minisatellites and assuming a model with an infinite number of minisatellites distributed over a finite number of clusters, the number of minisatellite clusters in the bovine genome has been estimated at ± 45, or 1.5 per chromosome.

We were able to demonstrate linkage between fingerprint bands and at least 3 out of 25 classic markers: transferring, phosphoglucomutase-1, protease inhibitor-2, and possibly thryroglobulin.

Two-locus analysis with the LINKAGE programs (Lathrop and Lalouel, 1984) was used to examine possible linkage of fingerprint bands with the MH and roan loci. In the MH analysis, the animals were classified as conventional, double-muscled, or of undetermined phenotype after visual examination. We allowed for visual misclassification of MH status by assuming variable penetrance for different genotypes. Moreover, plasma and red cell concentrations of creatine and creatinine, known to correlate well with muscular development with significantly different concentrations found for animals in the three different genotype classes (Hanset and Michaux, 1986), were incorporated into the linkage analysis assuming a generalized single-locus model, under the assumption that the observed phenotype distribution for a single variable is a mixture of normal distributions with common variance and different genotype means. Quantitative and qualitative (visual classification) data for the same locus were considered simultaneously. Means and variances for the four quantitative variables were estimated prior to linkage analysis from data in all pedigrees.

One band, EBA-3 detected with the human EFD-134.7 VNTR, gave lod scores > 4 (theta 0) when considering only visual classification. When including quantitative measurements of muscular development, the lod scores ranged from 2.84 to 4.10 (theta 0). EBA-3 was transmitted in coupling with the "mh" allele, without obligate recombinants. The EBA-3 is obviously a good candidate marker for the bovine MH gene. Cloning of the EBA-3 fingerprint band, which is in progress, and its use to generate a locus-specific polymorphism that can be traced in the same and other informative pedigrees should tell us whether EBA-3 is indeed a genetic marker for this intriguing gene.

No significant lod score was found between any fingerprint band and the "roan" locus. Exclusion analysis, however, demonstrated that 4.5 Morgans of the bovine genome had been explored with the fingerprint bands.

Therefore, despite the identification of a candidate marker for the "muscular hypertrophy" locus, this study indicates that DNA fingerprints may not be as useful for linkage studies as initially thought. The minisatellite sequences explored when performing DNA fingerprints in cattle are not randomly dispersed throughout the genome, but tend to be organized in clusters. Only part of the genome is thus accessible using this approach. This corroborates the mapping results of human VNTRs, but it seems to contradict the results obtained when performing linkage studies with mouse DNA fingerprints (Jeffreys et al., 1987). However, we believe that even assuming that clustering of minisatellites will be the rule, this approach remains valuable to trace specific genes in species where other marker systems are not yet available, especially because it is relatively cheap and rapid.

VNTR Markers in Domestic Animals

Although it is unlikely that complete genetic maps based only on VNTR type markers will ever be available, minisatellite sequences still remain an attractive source of genetic markers even if limited to defined regions of the genome. Several laboratories are now involved in the search for these markers in domestic species.

Screening purposely built libraries with probes known to reveal DNA fingerprints in cattle, we have isolated bovine minisatellites revealing highly polymorphic, locus-specific VNTR patterns when used as probes in high-stringency Southern blot hybridizations. More than 30 bovine VNTR markers have now been obtained in our laboratory (Georges et al., in preparation). One of these is shown in Figure 1. Heterozygosities ≥ 70% are common. These highly informative markers should allow us to efficiently cover a nonneglectable, although still to be determined, part of the bovine genome in linkage studies. Note that one of these VNTRs has been mapped by in situ hybridization to the

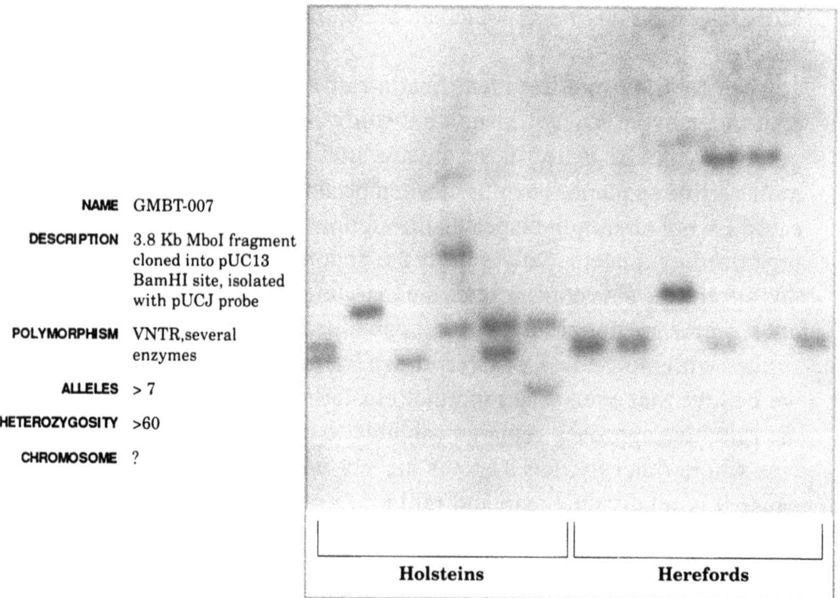

Figure 1 MboI restriction pattern obtained with the GMBT-007 cattle VNTR probe. At least seven different alleles could be recognized in a sample of 12 individuals; the heterozygosity was ≥ 60%.

telomere of chromosome 21 (Fries, personal communication), confirming Royle's (Royle et al., 1988) conclusion on the preferential proterminal location of human minisatellites.

Besides their use for linkage studies, we believe that animal VNTRs offer sufficient advantage to eventually replace or at least complement available blood group systems in paternity diagnosis. Overall, with the exception of the bovine B blood group system, VNTR are more polymorphic and thus more powerful than blood group systems. Combining five of our best cattle VNTR probes, characterized by heterozygosities $\geq 80\%$, the probabilities to miss wrong paternities would be $\pm 10^{-3}$ and 10^{-1}, when real and putative father are unrelated or father/son, respectively. With 10 probes, these probabilities become $\pm 10^{-6}$ and 10^{-2} respectively. DNA typing is not limited to blood samples but can be applied to any tissue sample containing nucleated cells, even perhaps hair roots [especially if applying polymerase chain reaction (PCR) methods]. Moreover, VNTR markers do not have the same disadvantages as multilocus DNA fingerprints: they are very reproducible, data banks could easily be set up, and results, even when generated in different laboratories could be easily compared. Finally, with continuously improving efficiency of DNA technology and automation being already available for several steps, costs of DNA tests become very competitive, even for animals.

To our knowledge, no sheep-specific VNTR has yet been identified. However, 75% of our bovine VNTRs recognize hypervariable minisatellites in sheep in a locus-specific way (Georges et al., in preparation). It is reasonable to speculate that the bovine probes explore the homologous locus in sheep. These minisatellite loci therefore antedate the divergence between cattle and sheep.

In pigs, Coppieters et al. (in press) serendipitously isolated a hypervariable minisatellite sequence, pS3, while looking for a Po-2 clone. The pS3 can be used both to detect a locus-specific VNTR in pigs and for multilocus DNA fingerprinting in pigs, horses, and rabbits.

CONCLUSIONS

As previously mentioned, animal geneticists, aware of the potentials offered by reverse genetics, were putting a lot of hope in hypervariable minisatellite sequences as a means to generate genetic maps of domestic animals in a reasonable time span. Thanks to the cross-species conservation of families of hypervariable minisatellites, minisatellites sequences could indeed be identified very quickly in several species, allowing both the establishment of multilocus DNA fingerprints and the isolation of species-and locus-specific VNTR markers. Both uses of minisatellite sequences should be very helpful in animal breeding, essentially for paternity diagnosis and linkage studies. However, apparent clustering of these sequences in defined regions of the genome allows only part of the genome to be explored when relying on them. In order to generate complete genetic maps, other highly polymorphic systems will have to be developed to complement the minisatellite markers.

Once again human geneticists are paving the way in the animal field, showing that, besides minisatellites, other hypervariable sequences might exist, apparently not confined to specific regions of the genome. These include microsatellites (Weber and May, 1989; Litt and Luty, 1989; Tautz, 1989) and polymorphism associated with the poly-A tracks characterizing in humans and rodents the so-called short interspersed retrotranscript repeats (SINES). Moreover, methods have been developed to detect a larger proportion of the existing genetic variation characterizing "normal" DNA sequences. These are based either on the melting behavior of DNA in denaturing gradient gel electrophoresis or on the susceptibility of single base mismatches to chemical or ribonuclease cleavage (Myers and Maniatis, 1986). These methods can now be combined with the PCR, increasing their power considerably (Sheffield et al., 1989).

Combining these different approaches should provide animal breeders fairly quickly with the genetic maps they have been waiting for, providing them with a tool to study a fascinating subject: *the molecular basis of artificial selection.*

REFERENCES

1. Ali S, Müller CR, Epplen JT. DNA fingerprinting by oligonucleotide probes specific for simple repeats. Hum Genet 1986; 74:239–243.
2. Armour JAL, Patel I, Thein SL, Fey MF, Jeffreys AJ. Analysis of somatic mutations at human minisatellite loci in tumors and cell lines. Genomics 1989; 4:328–334.

3. Beckmann JS, Soller M. Molecular markers in the improvement of farm animals. Biotechnology 1987; 5:573–576.
4. Bell GI, Selby MJ, Rutter WJ. The highly polymorphic region near the human insulin gene is composed of simple tandemly repeating sequences. Nature 1982; 295:31–35.
5. Bishop DT, Williamson JA, Skolnick MH. A model for restriction fragment length distributions. Am J Hum Genet 1983; 35:795–815.
6. Botstein D, White RL, Skolnick M, Dabis, RW. Construction of a genetic linkage map in man using restriction fragment length polymorphisms. Am J Hum Genet 1980; 32:314–331.
7. Burke T, Bruford MW. DNA fingerprinting in birds. Nature 1987; 327:149–152.
8. Capon DJ, Chen EY, Levinson AD, Seeburg PH, Goeddel DV. Complete nucleotide sequence of the T24 human bladder carcinoma oncogene and its normal homologue. Nature 1983; 302:33–37.
9. Cohen JB, Walter MV, Levinson AD. A repetitive sequence element 3′ of the human c-Ha-ras1 gene has enhancer activity. J Cell Physiol 1987; 5(Suppl):75–81.
10. Coppieters W, Van De Weghe A, Depicker, A, Bouquet Y, Van Zeveren A. A hypervariable pig DNA fragment. Anim Genet (in press).
11. Dallas JF. Detection of DNA "fingerprints" of cultivated rice by hybridization with a human minisatellite DNA probe. Proc Natl Acad Sci USA 1988; 85:6831–6835.
12. Farber CM, Georges M, Debock G, Verhest A, Simon P, Verschraegen-Spae M. Demonstration of spontaneous XX/XY chimerism by DNA fingerprinting. Hum Genet 1989; 182:1–3.
13. Fowler C, Drinkwater R, Burgoyne L, Skinner J. Hypervariable lengths of human DNA associated with a human satellite III sequence found in the 3.4 kb Y-specific fragment. Nucl Acids Res 1987; 15:3929.
14. Fries R, Beckmann JS, Georges M, Soller M, Womack J. The bovine gene map. Anim Genet 1989; 20:3–29.
15. Goerges M, Lequarré AS, Hanset R, Vassart G. Genetic variation of the bovine thyroglobulin locus studied at the DNA level. Anim Genet 1987a; 18:41–50.
16. Georges M, Hilbert P, Lequarré A, Leclerc V, Hanset R, Vassart G. Use of DNA bar codes to resolve a canine paternity dispute J Am Vet Med Assoc 1988b; 193:9.
17. Georges M, Lathrop M, Hilbert P, Marcotte A, Schwers A, Swillens S, Vassart G, Hanset R. On the use of DNA fingerprints for linkage studies in cattle. Genomics 1990 6:461–474.

18. Goodbourn SEY, Higgs DR, Clegg JB, Weatherall DJ. Molecular basis of length polymorphism in the human α-globin gene complex. Proc Natl Acad Sci USA 1983; 80:5022-5026.
19. Gyllensten UB, Jakobsson S, Temrin H. Wilson AC. Nucleotide sequence and genomic organization of bird minisatellites. Nucl Acids Res 1989; 17:2203-2214.
20. Hanset R. Recherches sur la white heifer disease et son determinisme genetique Brossells: Comtes-Rendus de Recherche—IRSIA nr. 33. 1965:177.
21. Hanset R. Coat colour inheritance in the Belgian white and blue cattle breed. Gén Sél Evol 1985; 17(4):443-458.
22. Hanset R, Michaux C. On the genetic determinism of muscular hypertrophy in the Belgian white and blue cattle breed. I. Experimental data. Gén Sél Evol 1985a: 17(3):359-368.
23. Hanset R, Michaux C. On the genetic determinism of muscular hypertrophy in the Belgian white and blue cattle breed. II. Population data. Gén Sél Evol 1985b; 17(3):369-386.
24. Hanset R, Michaux C. Characterization of biological types of cattle by the blood levels of creatine and creatinine. J Anim Breed Genet 1986; 103:227-240.
25. Hilbert P, Marcotte A, Schwers A, Hanset R, Vassart G, Georges M. Analysis of genetic variation in the Belgian blue cattle breed using DNA sequence polymorphism at the growth hormone, low density lipoprotein receptor, α-subunit of glycoprotein hormones and thyroglobulin loci. Anim Genet 1989; 20:385-394.
26. Hill AVS, Jeffreys AJ. Use of minisatellite DNA probes for determination of twin zygosity at birth. Lancet 1985; 1:1394-1395.
27. Jarman AP, Higgs DR. A new hypervariable marker for the human α-globin gene cluster. Am J Hum Genet 1988; 43:249-256.
28. Jarman AP, Nicholls RD, Weatherall DJ, Clegg JB, Higgs DR. Molecular characterization of a hypervariable region downstream of the human α-globin gene cluster. EMBO J 1986; 5:1857-1863.
29. Jeffreys AJ, Morton DB. DNA fingerprints of dogs and cats. Anim Genet 1987; 18:1-15.
30. Jeffreys AJ, Wilson V. Individual-specific "fingerprints" of human DNA. Nature 1985b; 316:76-79.
31. Jeffreys AJ, Wilson V, Thein SL. Hypervariable "minisatellite" regions in human DNA. Nature 1985a; 314:67-73.

32. Jeffreys AJ, Wilson V, Thein SL, Weatherall DJ, Ponder BAJ. DNA "fingerprints" and segregation analysis of multiple markers in human pedigrees. Am J Hum Genet 1986; 39:11-24.
33. Jeffreys AJ, Wilson V, Kelly R, Taylor BA, Bulfield G. Mouse DNA "fingerprints": analysis of chromosome localization and germ-line stability of hypervariable loci in recombinant inbred strains. Nucl Acids REs 1987; 15:2823-2836.
34. Jeffreys AJ, Royle NJ, Wilson V, Wong Z. Spontaneous mutation rates to new length alleles at tandem-repetitive hypervariable loci in human DNA. Nature 1988; 332:278-280.
35. Knott TJ, Wallis SC, Pease RJ, Powell LM, Scott J. A hypervariable region 3′ to the human apolipoprotein B gene. Nucl Acids REs 1986; 14:9215-9216.
36. Kuhnlein U, Dawe Y, Zadworny D, Gavora JS. DNA fingerprinting: a tool for determining genetic distances between strains of poultry. Theor Appl Genet 1989; 77:669-672.
37. Lathrop GM, Lalouel JM. Easy calculations of lod scores and genetic risks on small computers. Am J Hum Genet 1984; 36:460-465.
38. Lathrop M, Nakamura Y, Cartwright P, O'Connell P, Leppert M, Jones C, Tateishi H, Bragg T, Lalouel J, White R. A mapped set of genetic markers for human chromosome 10. Genomics 1988a; 2:157- 164.
39. Lathrop M, Nakamura Y, O'Connell P, Leppert M, Woodward S, Lalouel J, White R. A mapped set of genetic markers for human chromosome 9. Genomics 1988b; 3:361-366.
40. Litt M, Luty JA. A hypervariable microsatellite revealed by in vitro amplification of a dinucleotide repeat within the cardiac muscle actin gene. Am J Hum Genet 1989; 44:397-401.
41. Martin B, Nienhuis J, King G, Schaefer A. Restriction fragment length polymorphisms associated with water use efficiency in tomato. Science 1989; 243:1725-1727.
42. Morton DB, Yaxley RE, Patel I, Jeffreys AJ, Howes SJ, Debenham PG. Use of DNA fingerprint analysis in identification of the sire. VEt Rec 1987; 121:592-593.
43. Myers RM, Maniatis T. Cold Spring Harbor Symp Quant Biol 1986; 51:271-284.
44. Nakamura Y, Leppert M, O'Connell P, Wolff R, Holm T, Culver M, Martin C, Fujimoto E. Hoff M, Kumlin E. White R. Variable number of tandem repet (VNTR) markers for human gene mapping. Science 1987; 235:1616-1622.

45. Nakamura Y, Lathrop M, O'Connell P, Leppert M, Lalouel JM, White R. A mapped set of DNA markers for human chromosome 15. Genomics 1988a; 3:342-346.
46. Nakamura Y, Lathrop M, O'Connell P, Leppert M, Barker D, Wright E, Skolnick M, Kondoleon S, Litt M, Lalouel JM, White R. A mapped set of DNA markers for human chromosome 17. Genomics 1988b; 2:302-309.
47. Nakamura Y, Lathrop M, O'Connell P, Leppert M, Lalouel JM, White R. A primary map of ten DNA markers and two serological markers for human chromosome 19. Genomics 1988c; 3:67-71.
48. O'Connell P, Lathrop GM, Law M, Leppert ML, Nakamura Y, Hoff M, Kumlin E, Thomas W, Elsner T, Ballard L, Goodman P, Azen E, Sadler J, Cai G, Lalouel JM, White R. A primary genetic linkage map for human chromosome 12. Genomics 1987; 1:93-102.
49. O'Connell P, Lathrop GM, Nakamura Y, Leppert ML, Ardinger RH, Murray JL, Lalouel JM, White R. Twenty-eight loci form a continuous linkage map of markers for human chromosome 1. Genomics 1989; 4:12-20.
50. O'Hara PJ, Grant FJ. The human factor VII gene is polymorphi due to variation in repeat copy number in a minisatellite. Gene 1988; 66:147-158.
51. Orkin SH. Reverse genetics and human disease. Cell 1986; 47:845-850.
52. Paterson AH, Lander ES, Hewitt JD, Peterson S, Lincoln SE, Tanksley SD. Resolution of quantitative traits into mendelian factors by using a complete linkage map of restriction fragment length polymorphisms. Nature 1988; 335:721-726.
53. Perret J, Shia Y, Fries R, Vassart G, Georges M. A polymorphic satellite sequence maps to the pericentric region of the bovine Y chromosome. Genomics (in press).
54. Rogstad SH, Patton JC II, Schaal BA. M13 repeat probe detects DNA minisatellite-like sequences in gymnosperms and angiosperms. Proc Natl Acad Sci USA 1988; 85:9176-9178.
55. Royle NJ, Clarkson RE, Wong Z, Jeffreys AJ. Clustering of hypervariable minisatellites in the proterminal regions of human autosomes. Genomics 1988; 3:352-360.
56. Ryskov AP, Jincharadze AG, Prosnyak MI, Ivanov PL, Limborska SA. M13 phage DNA as a universal marker for DNA fingerprinting of animals, plants and microorganisms. FEBS lett 1988; 233(2):388-389.
57. Schafer R, Zischler H, Birsner U, Becker A, Epplen JT. Optimized oligonucleotide probes for DNA fingerprinting. Electrophoresis 1988; 9:369-374.

58. Sheffield VC, Cox DR, Lerman LS, Myers RM. Attachment of a 40-base pair G+C-rich sequence (GC-clamp) to genomic DNA fragments by the polymerase chain reaction results in improved detection of single-base changes. Proc Natl Acad Sci USA 1989; 86:232-236.
59. Sen D, Gilbert W. Formation of parallel four-stranded complexes by guanine-rich motifs in DNA and implications for meiosis. Nature 1988; 334:364-366.
60. Simmloer MC, Johnsson C, Petit C, Rouyer F, Vergnaud G, Weissenbach J. Two highly polymorphic minisatellites from the pseudo-autosomal region of the human sex chromosomes. EMBO J 1987; 6(4):963-969.
61. Smith C, Simpson SP. The use of genetic polymorphism in livestock improvement. J Anim Breed Genet 1986; 103:205-217.
62. Soller M, Beckmanm JS. Restriction fragment length polymorphisms and genetic improvement. 2nd World Congress on Genetics Applied to Livestock Production. Madrid 1982; 6:396-404.
63. Stocker NG, Cheah KSE, Griffin JR, Pope RM, Solomon E. A highly polymorphic region 3′ to the human type II collagen gene. Nucl Acids Res 1985; 13:4613-4622.
64. Swallow DM, Gendler S, Griffiths B, Corney G, Taylor-Papadimitriou J, Bramwell ME. The human tumour-associated etithelial micins are coded by an expressed hypervariable gene locus PUM. Nature 1987; 328:82-84.
65. Tautz D. Hypervariability of simple sequences as a general source for polymorphic DNA markers. Nucl Acids Res 1989; 17(16):6463-6471.
66. Thein SL, Jeffreys AJ, Blacklock HA Identification of post-transplant cell population by DNA fingerprint analysis. Lancet 1986; 2:37.
67. Vassart G, Georges M, Monsieur R, Brocas H, Lequarre AS, Christophe D. A sequence in M13 phage detects hypervariable minisatellites in human and animal DNA. Science 1987; 235:683-684.
68. Wever JL, May PE. Abundant class of human DNA polymorphisms which can be typed using the polymerase chain reaction. Am J Hum Genet 1989; 44:388-396.
69. Wetton JH, Carter RE, Parkin DT, Walters D. Demographic study of a wild house sparrow population by DNA fingerprinting. Nature 1987; 327:147-149.
70. Wiggs J, Nordenskjold M, Yandell D, Rapaport J, Grondin V, Janson M, Werelius B, Petersen R, Craft AL, Riedel K, Liberfarb R, Walton D, Wilson W, Dryja TP. Prediction of the risk of hereditary retinoblastoma,

using DNA polymorphisms within the retinoblastoma gene. N Engl J Med 1988; 318(3):151-157.

71. Wolff RK, Nakamura Y, White R. Molecular characterization of a spontaneously generated new allele at a VNTR locus: no exchange of flanking DNA sequence. Genomics 1988; 3:347-351.
72. Wolff RK, Plaetke R, Jeffreys AJ, White R. Unequal crossingover between homologous chromosomes is not the major mechanism involved in the generation of new alleles at FNTR loci. 1989.
73. Wong Z, Wilson V, Jeffreys AJ, Thein SL. Cloning a selected fragment from a human DNA "fingerprint": isolation of an extrtemely polymorphic minisatellite. Nucl Acids Res 1986; 14:4605-4616.
74. Wong A, Wilson V, Patel I, Povey S, Jeffreys AJ. Characterization of a panel of highly variable minisatellites cloned from human DNA. Ann Hum Genet 1987; 51:269-288.
75. Wyman A, White R. A highly polymorphic locus in human DNA. Proc Natl Acad Sci USA 1980; 77:6754-6758.?

6

In Situ Hybridization and High–Resolution Banding of Chromosomes

C. Larry Chrisman,† Glenn P. Briley, and Geoffrey C. Waldbieser

Purdue University
West Lafayette, Indiana

INTRODUCTION

In situ hybridization of nucleic acids to chromosomes has become a widely used method of gene mapping in a variety of species. Hsu (1979) credits Gall and Pardue (1969) for developing this technique, which combines the fields of molecular genetics and cytogenetics. Since single–copy genes were first localized in 1981 (Harper and Saunders, 1981; Gerhard et al., 1981), human gene mapping and, more recently, domestic animal gene mapping have been advanced through the precise localization provided by this method. Gene localization to a specific chromosomal area is achieved by incubating denatured chromosomal DNA with a labeled nucleic acid probe that contains sequence homology with the gene of interest. Chromosomal sites that have annealed with the probes are identified via differential staining of each chromosome (chromosome banding). This method depends not only on successful hybridization and probe detection, but also on identifiable chromosome banding patterns. Although the method of Harper and Saun-

†Deceased.

ders (1981) has become the standard approach, products and techniques are continually being developed to improve the sensitivity of in situ hybridization.

Applications of gene localization to chromosomes are numerous. The identification of members of syntenic groups of genes may yield information about both evolution and gene function. Identification and localization of genes with beneficial or detrimental major effects will enhance selection programs aimed at improvement of production efficiency in species of agricultural importance. By localizing integration sites of foreign genes in transgenic animals, the importance of insertion site to gene function, insertional mutagenesis, and perhaps the nature of the insertion site itself can be determined. Localization of genes to sites involved in chromosome translocations may aid in revealing the environmental mutagenic events leading to the formation of fusion genes involved in cancer. This information might also be useful for cancer therapy, such as in the preparation of pharmaceuticals that interact specifically with the recombinant genes or gene product resulting from the translocation event.

PREPARATION OF CHROMOSOMES

When preparing metaphase chromosomes for light microscopic evaluation, one requires a source of dividing cells, the arrest of mitotic division, and preservation of the chromosomes in a structural form that is meaningful and repeatable. Most cytogenetic methods utilize chromosomes obtained from prophase, prometaphase, and metaphase cells preserved by methanol-acetic acid fixation. These cells may be obtained directly from tissues that contain large numbers of rapidly dividing cells, such as hemopoietic, germinal, or neoplastic tissues. Cells from other tissues may be grown in long-term culture to amass enough dividing cells for efficient preparation of chromosomes. For most routine analyses, chromosomes are prepared from blood lymphocytes that have undergone mitotic stimulation in short-term culture.

Lymphocyte Culture

The method commonly used to obtain mitotic cells from most species is lymphocyte culture (Priest, 1977; Rooney and Czepulkowski, 1986).

This technique takes advantage of mitogenic compounds that stimulate peripheral blood lymphocytes to dedifferentiate and undergo clonal division. Such compounds include phytohemagglutinin, pokeweed mitogen, concanavalin A, and certain lipopolysaccharides.

A typical lymphocyte culture contains 5-10 ml nutrient media, buffered with bicarbonate or HEPES, in a sterile flask or glass bottle. The media is supplemented with serum, antibiotics, and appropriate mitogen. Often, fresh L-glutamine is added to the media since this necessary amino acid is oxidized after storage. Whole blood or leukocyte-enriched plasma is added to the culture, which is then incubated for 48-72 h. Mammalian cultures are usually incubated at 37°C, whereas cultures from fish and other poikilotherms may require alternative incubation temperatures (Wolters et al., 1981).

Chromosomes exhibit a reproducible structural pattern when obtained from prophase, prometaphase, and metaphase cells. Since chromosome length decreases as mitosis progresses, prophase and prometaphase cells are required for high-resolution analyses. One may amplify the number of cells at the metaphase stage by blocking karyokinesis via spindle poisons (colchicine or vinblastine). This treatment immobilizes cell division at metaphase, and mitotic cells collect at this stage with time.

Alternately, one may synchronize cell division and increase the number of prophase and prometaphase cells in the culture. Chemicals that block DNA synthesis, such as fluorouracil, methotrexate, and high levels of thymidine or 5-bromodeoxyuridine (BrdU), can be added to the culture to collect cells at the synthesis phase of the cell cycle (Priest, 1977; Rooney and Czepulkowski, 1986). Release of the block provides a number of cells that progress through the cell cycle in synchrony. Empirical methods may be used to determine the time at which these cells enter prophase, prometaphase, and metaphase. Again, spindle poisons can be added to inhibit karyokinesis and increase yield.

In order to harvest chromosomes, the cultured cells are decanted or washed into a centrifuge tube and spun into a pellet. The media is removed and the cells are exposed to a hypotonic solution, usually 0.56% potassium chloride, although 1.0% trisodium citrate can be used. Although time of hypotonic treatment varies between laboratories, it is usually applied from 4 to 30 min. The cells are preserved by removing the hypotonic solution and adding a fixative (3 parts methanol : 1 part acetic acid), and are then washed several times with fixative to remove cellular debris. Osmotic swelling during hypotonic treatment assists in spreading of the chromosomes when the cell suspension is dropped onto a glass slide.

Tissue Culture

In order to prepare chromosomes from tissues with low mitotic activity, long-term cell and tissue culture is often employed. Long-term culture requires more expertise and equipment than lymphocyte culture. However, long-term cell cultures may be necessary in some species where high-quality chromosome preparations from lymphocyte cultures are difficult to obtain. The cells are maintained in nutrient media to 60–70% confluence, and then harvested while still in mitotic growth phase. Synchronization methods used in lymphocyte cultures can also be applied to long-term cultures to increase the yield of metaphase cells (Freshney, 1983). At harvest, cells attached to the culture flask are dislodged by either mechanical (shaking or scraping) or enzymatic (trypsin or collagenase) methods, and then treated as described above.

Solid Tissues

Mitotic and meiotic cells are often obtained from solid tissues (Priest, 1977), thereby eliminating the need for tissue culture. In healthy animals, the hemopoietic tissues, such as bone marrow and spleen in mammals or kidney tissue in fish, often supply large numbers of mitotic cells amenable for chromosome preparation. Testes and postovulatory ova are used as sources of meiotic cells. Preparation from these tissues usually requires only that the cells be well dispersed from connective tissue and support cells. Thus a mechanical or chemical separation is employed to suspend the mitotic cells in an isotonic medium before hypotonic treatment. Alternatively, the cells may be disaggregated directly in hypotonic solution. Fixation and slide preparation follows as described above.

PREPARATION OF NUCLEIC ACID PROBES

While single-stranded DNA and RNA may be used as probes for gene localization, double-stranded DNA is most commonly used for in situ hybridization to chromosomes. Whether a sequence of genomic DNA or cDNA, this probe can be inserted into a plasmid for amplification and storage. Gene-specific sequences can be excised from the plasmid for use as a probe, or plasmid DNA can be included in the probe to enhance network formation (Wahl et al., 1979). After denaturation, complimentary vector sequences can reanneal to form multiprobe networks of single-stranded DNA. This network concentrates the number of probe molecules at the chromosomal hybridization site.

Nucleic acid probes must be radioactively or chemically labeled in order to detect the site of hybridization. Nick translation (Rigby et al., 1977), which utilizes the excision repair qualities of DNA polymerase I, and random primer labeling (Feinberg and Vogelstein, 1983, 1984), which employs the DNA chain extension properties of DNA polymerase I, are used to incorporate labeled nucleotides into the probe sequence. Polymerase chain reaction (PCR) is also a highly efficient method of labeling DNA (Schowalter and Sommer, 1989). This technique, which uses primer-directed amplification of a DNA sequence, can produce large amounts (10- to 10,000-fold) of labeled probe DNA (Innis et al., 1990).

Radiolabeling of DNA probes entails the incorporation of radioactive nucleotides into the DNA sequence. The choice of isotope for radiolabeling depends on the desired precision of the hybridization signal and time allotment for the experiment. Isotopes of high specific activity, such as ^{32}P and ^{125}I, allow a short exposure time during autoradiography but imprecise localization due to overexposure of the photographic emulsion. High resolution may be achieved by labeling with ^{3}H (tritium), which has an average path length of only 1 μm in emulsion. Tritiated probes require lengthy exposure times (e.g., 2–4 weeks), however, because the specific activity of this isotope is low. ^{35}S offers a compromise between sensitivity and resolution since it emits particles of higher energy than ^{3}H, but results in less precise localization.

Alternatively, nonisotopic, chemical labeling methods utilize biotin (Langer et al., 1981) or its photoreactive analog, photobiotin (Forster et al., 1985). Biotinylated dUTP is incorporated into the probe sequence whereas photobiotin links to DNA when activated by ultraviolet light. The success of this system is due to the strong interaction of biotin or photobiotin with avidin or streptavidin, two biotin-binding proteins that can be linked to a variety of compounds to facilitate detection of the probe. Compared to radiolabeled probes, the biotinylated probes do not require autoradiography, and signals can be detected soon (4 h) after hybridization.

HYBRIDIZATION

The hybridization procedure using a radiolabeled or biotinylated probe is a combination of several steps, each of which is important for obtaining a localized signal with little background. These steps include ribonuclease treatment, denaturation of chromosomal DNA, hybridization, posthybridization washing, and signal detection (Modi et al., 1987; Naylor et al., 1987; MacGregor and Varley, 1988).

Slides containing metaphase chromosomes are incubated in a ribonuclease solution to destroy RNA and prevent probe hybridization to nuclease nonchromosomal nucleic acids. Before this treatment, the enzyme solution is boiled to remove any contaminating deoxyribonucleases that may degrade the chromosomal DNA. After the enzyme treatment, the slides are washed in 2 X SSC (1 X SSC = 0.15 M sodium chloride and 0.015 M sodium citrate) and dehydrated in an ethanol series.

Chromosomal DNA may be denatured either by heating in a formamide solution or by incubating at high pH. Formamide lowers the melting point of DNA, enabling heat denaturation at a lower temperature. With a solution of 70% formamide/2 X SSC, the melting temperature is decreased from approximately 90°C to 70°C. Alternatively, the slides may be incubated at 37°C in a 1 M solution of NaOH. Denaturation at lower temperatures may prevent loss of DNA from the slide. Following denaturation, the slides are washed in a cold (4°C) ethanol series and air-dried.

The concentration of labeled probe for in situ hybridization is generally 1-50 ng/ml. The probe is added to the hybridization buffer, containing 50% formamide for efficient hybridization at 37°C. Dextran sulfate, also included in the buffer, increases the effective concentration of the probe and promotes the formation of DNA networks (Wahl et al., 1979). Low-molecular-weight (carrier) DNA, such as sonicated salmon sperm or calf thymus DNA, is added to the buffer at a concentration of 500-1000X that of the probe to block nonspecific DNA hybridization. The probe is denatured by heating the hybridization mixture to 95°C. After cooling on ice, the hybridization buffer (30-50 μl) is applied to the slide, and then a coverslip is placed on the slide and sealed with rubber cement. Slides are incubated overnight at 37°C in a humidified environment. After hybridization, the slide is thoroughly washed to retain only strongly bound probes.

Autoradiography is necessary to visualize the signal produced by the radioactive probe. In a darkroom, the slides are dipped into a liquid emulsion (such as Kodak NTB2) that has been diluted with water and warmed to 42-45°C. After the slides have air-dried, they are placed in light-tight boxes containing desiccant (Drierite or silica gel) and stored at 4°C for 2-4 weeks (tritium-labeled probes). Additional slides are often included as "test" slides and are developed first to check signal strength and background. The slides are developed in freshly prepared Kodak D-19 or Dektol, washed with distilled water and fixed in Kodak General Fixer. All solutions, from developer to final wash, are kept at 15-20°C to prevent wrinkling of the emulsion. After air drying, emulsion on the back of each slide should be removed by scraping with a razor blade. Details of tritium autoradiography may be found in MacGregor and Varley (1988).

Current nonisotopic detection systems exploit the high affinity of avidin or streptavidin for biotin. Strepavidin, isolated from *Strepomyces avidinii* is preferred over egg-white avidin, which exhibits higher nonspecific binding that streptavidin. Strepavidin can be conjugated with an enzyme, such as horseradish peroxidase or alkaline phosphatase, which converts a chromogenic substrate into a localized precipitate. Detection is performed immediately after posthybridization washes, and it is important to keep the slides moist during the procedure (e.g., the ethanol dehydration step is omitted). The slides are first incubated in a solution containing conjugated streptavidin, which binds to the biotin-labeled DNA probe. After washing, the chromogenic substrate is applied and precipitation occurs at the hybridization site. The streptavidin-horseradish peroxidase complex, when reacted with diaminobenzidine (DAB), produces a brown DAB precipitate (Crolla and Llerena, 1988). A blue precipitate is produced by incubation of alkaline phosphatase-streptavidin with 5-bromo-4-chloro-3-indolyl phosphate (BCIP) and nitro blue tetrazolium (NBT) (Garson et al., 1987). In addition to these direct methods, antibodies can be used to amplify the biotinylated signal before detection, and silver precipitation has been used to enhance the DAB/peroxidase signal (Burns et al., 1985; Bhatt et al., 1988). With each method, signal detection is rapid and the slides may be viewed with light microscopy after appropriate staining.

Biotinylated DNA probes are also detected with a fluorescent signal. Streptavidin conjugate with a fluorescent compound, such as fluorescein or rhodamine, can be used for direct probe detection. Antibody amplification has been used in fluorescein-avidin detection (Cherif et al., 1989), producing a yellow-green signal against red chromosomes when viewed with fluorescence microscopy.

Metaphase chromosomes are viewed under the light microscope after the slides have been stained in a buffered Giemsa solution. When analyzing actual hybridization events, one must decide whether to count grains that touch the edge of the chromosomes as well as those that lie on chromosomes. For single-copy genes, 50–100 metaphase figures should be examined in order to plot grain distribution. Grain positions are recorded on an ideogram of the banded karyotype so that a significant hybridization site can be easily identified. Usually, the greatest number of grains in a single position is approximately 20% of the total number of grains scored. Although this percentage may seem quite small, a significant secondary localization is usually not found. One can also use chi-square analyses to compare actual grain distribution to the distribution expected if grain deposition was random. Calculation of random grain distribution must take into account the relative amount of genomic DNA per chromosome (Gerhard et al., 1981).

HIGH-RESOLUTION CHROMOSOME BANDING WITH IN SITU HYBRIDIZATION

The subchromosomal localization of the hybridization signal requires structural markers along the length of the chromosome. Several techniques have been developed that produce a differential, reproducible staining pattern dependent on individual chromosomal architecture. Although several chromosome "banding" techniques have been developed, this section will concentrate on Q-, G-, and R-bands, which are most widely used in animal cytogenetics.

Chromosome banding techniques for in situ hybridization fall into two main categories, prebanding and postbanding. Prebanding requires the staining and photography of suitable metaphase chromosomes prior to denaturation and is indicated when the staining procedure does not produce high-quality bands after hybridization. This procedure provides sharp bands, as are normally produced for routinely banded chromosomes, but may limit the number of analyzable metaphase figures after in situ hybridization. Postbanding simply involves the banding of hybridized chromosomes after the in situ procedure. With postbanding, more metaphase figures can be scored, but the analyses are complicated if the banding procedure fails. Also, the photographic emulsion can drastically interfere with postbanding procedures. One should first photograph conventionally stained metaphase figures in case the probe is removed during the banding procedure.

Quinacrine, or Q-, banding utilizes the fluorescent dyes quinacrine dihydrochloride or Hoechst 33258, which preferentially intercalate into A+T- rich regions of chromosomal DNA (Caspersson et al., 1968). Q-banding produces a rapid, dependable banding pattern, but the chromosomes must be viewed via fluorescence microscopy and bands are temporary since the fluorescence quenches. Owing to low contrast, Q-banding may not be the optimal procedure for high resolution, but it is still employed as a reliable prehybridization banding technique for identifying major bands (Fries et al., 1988).

Giemsa (G-) banding techniques are probably the most widely used with in situ hybridization. G-banding employs the positive (dark) staining of late S- phase-replicating DNA, which is abundant in A+T sequences, and therefore highlights possible structural differences in the chromosomal DNA. Structural G-bands, which correlate with Q-bands, may be produced by staining with Wright's stain (Chandler and Yunis, 1978; Harper and Saunders, 1981; Cannizzaro and Emanuel, 1984) or by trypsin digestion followed by Giemsa staining (Seabright, 1971; Popescu et al., 1985). High-resolution, replicative G-bands are produced by incorporating BrdU during the early S-phase of the cell cycle into A+T regions of early replicating (G+C rich) DNA. The cell cycle is blocked during mid-S-phase and thymidine is incorporated during late S-phase after release of the block. The harvesting procedure is timed to yield prometaphase chromosomes. After incubation with Hoechst 33258, which binds strongly to A+BrdU sequences, the chromosomes are exposed to ultraviolet light. This treatment leads to preferential degradation of stainable material in BrdU-substituted regions (G-C rich) so that A+T-rich regions are preferentially stained (Perry and Wolff, 1974; Modi et al., 1987). Although this procedure leads to high-quality G-bands on elongated chromosomes, the resolution will deteriorate if photographic emulsion is used.

In contrast to Q- or G-banding, reverse (R-) banding leads to positive staining of early replicative (G+C rich) regions of chromosomal DNA. Two advantages of R-banding over Q- and G-banding, positive staining of telomeres and the ease of identification of centromeres, aid in chromosome identification. Also, many translocation breakpoints occur in G+C-rich DNA, so one can more easily localize these sites on R-banded chromosomes.

Structural R-banding, created by incubation in a hot salt solution followed by Giemsa staining (Dutrillaux and Lejeune, 1971), has been used as a prebanding procedure for in situ hybridization (Bernheim et al., 1984). High-resolution R-banding involves cell synchronization and BrdU incorporation into late replicative (A+T rich) DNA (Rønne, 1989; Drouin et al., 1989). After release of the methotrexate or florouracil block, BrdU is added to the culture. Subsequent addition of Hoechst 33258 to the culture leads to preferential intercalation of this fluorochrome into BrdU-substituted regions. The presence of Hoechst 33258 in the A+T-rich regions inhibits chromosome condensation as mitosis progresses. Fixed chromosomes are subjected to ultraviolet light, which preferentially degrades the BrdU-substituted (A+T rich)

DNA regions. The staining of G+C-rich regions produces the R-banding pattern (Perry and Wolff, 1974; Marlhens et al., 1987).

As with any laboratory methodology, in situ hybridization procedures vary between research groups. Often, the experience of the cytogenetics group involved will dictate the type of chromosome banding and whether to band pre- or posthybridization. Therefore, some laboratories select Q-banding for its relative simplicity and repeatability, while others choose G-banding due to increased (compared to Q-banding) resolution between subchromosomal regions. Still another group would insist on using R-banding of elongated chromosomes to provide maximum resolution. Choice of banding technique is also influenced by the precision of individual chromosome identification in the species of interest. Any conference devoted to chromosome studies will find opponents and proponents of each method discussed. Obviously, each research group will select a method that works best for it. However, correlations between Q-, G-, and R-banded chromosome preparations are not perfect. As more loci are identified by in situ hybridization, some standardization in reporting band location will be necessary.

IN MEMORIAM

Dr. Larry Chrisman passed away on November 22, 1990, at the age of 49. Dr. Chrisman enthusiastically promoted the effort to map the genomes of domestic animals, and he continually sought applications of cytogenetics and biotechnology to improve animal production efficiency. His early research explored the effects of hyperthermic stress on fertility in mammals. Dr. Chrisman made a major effort to incorporate genetic engineering into aquaculture research and was one of the first to study ploidy manipulation in warmwater fish. He participated in numerous symposia with state and federal agencies on the development of guidelines and policies on research funding, regulation of transgenic animal research, and patenting issues in biotechnology. The many scientists he trained have embraced his pioneering spirit and view their research not as an end in itself, but, as Dr. Chrisman would say, "part of the big picture."

REFERENCES

Bernheim A, Berger R, Szabo P. Localization of actin-related sequences by in situ hybridization to R-banded human chromosomes. Chromosoma (Berl) 1984; 89:163-167.

Bhatt B, Burns J, Flannery D, McGee JO'D. Direct visualization of single copy genes on banded metaphase chromosomes by nonisotopic in situ hybridization. Nucl Acids Res 1988; 16:3951-3961.

Burns J, Chan VTW, Jonasson JA, Fleming KA, Taylor S, McGee JO'D. Sensitive system for visualizing biotinylated DNA probes hybridized in situ: rapid sex determination of intact cells. J Clin Pathol 1985; 38:1085-1092.

Cannizzaro LA, Emanuel BS. An improved method for G-banding chromosomes after in situ hybridization. Cytogenet Cell Genet 1984; 38:308-309.

Caspersson T, Farber S, Foley GE, Kudynowski J, Modest EJ, Simmonsson E, Wagh U, Zech L. Chemical differentiation along metaphase chromosomes. Exp Cell Res 1968; 49:219-222.

Chandler ME, Yunis JJ. A high resolution in situ hybridization technique for the direct visualization of labeled G-banded early metaphase and prophase chromosomes. Cytogenet Cell Genet 1978; 22:352-356.

Cheriff D, Bernard O, Berger R. Detection of single-copy genes by nonisotopic in situ hybridization of human chromosomes. Hum Genet 1989; 81:358-362.

Crolla JA, Llerena JC Jr. A mosaic 45,X/46,X,r(?) karyotype investigated with X and Y centromere- specific probes using a non-autoradiographic in situ hybridization technique. Hum Genet 1988; 81:81-84.

Drouin R, Lemieux N, Richer C. High resolution R-banding at the 1250 -band level. 1. Technical considerations on cell synchronization and R-banding (RHG and RGB). Cytobios 1989; 56:107-125.

Dutrillaux B, Lejeune J. Sur une nouvelle technique d'analyse du caryotype humaine. C R Acad Sci 1971; 272:2638.

Feinberg AP, Vogelstein B. A technique for radiolabeling DNA restriction endonuclease fragments to high specific activity. Analyt Biochem 1983; 132:6-13.

Feinberg AP, Vogelstein B. A technique for radiolabeling DNA restriction endonuclease fragments to high specific activity—addendum. Analyt Biochem 1984; 137:266-267.

Forster AC, McInnes JL, Skingle DC, Symons RH. Non-radioactive hybridization probes prepared by the chemical labelling of DNA and RNA with a novel reagent, photobiotin. Nucl Acids Res 1985; 13:745-761.

Freshney RI. Culture of animal cells. New York: Alan R. Liss, 1983.

Fries R, Hediger R, Stranzinger G. The loci for parathyroid hormone and B-globin are closely linked and map to chromosome 15 in cattle. Genomics 1988; 3:302-307.

Gall JG, Pardue MC. Formation and detection of RNA-DNA hybrid molecules in cytological preparations. Proc Natl Acad Sci USA 1969; 63:378-383.

Garson JA, van den Berghe JA, Kemshead JT. Novel non-isotopic technique detects small (1 kb) unique sequences in routinely G-banded human chromosomes: fine mapping of N-myc and B-NGF genes. Nucl Acids Res 1987; 15:4761-4770.

Gerhard DS, Kawasaki ES, Bancroft FC, Szabo P. Localization of a unique gene by direct hybridization in situ. Proc Natl Acad Sci USA 1981; 78:3755-3759.

Harper ME, Saunders GF. Localization of single copy DNA sequences on G-banded human chromosomes by in situ hybridization. Chromosoma (Berl) 1981; 83:431- 439.

Hsu TC. Human and mammalian cytogenetics. An historical perspective. New York: Springer-Verlag, 1979.

Innis MA, Gelfand DH, Sninsky JJ, White TJ, eds. PCR protocols. A Guide to methods and applications. San Diego: Academic Press, 1990.

Langer PR, Waldrop AA, Ward DC. Enzymatic synthesis of biotin-labeled polynucleotides: novel nucleic acid affinity probes. Proc Natl Acad Sci USA 1981; 78:6633-6637.

MacGregor H, Varley J. Working with animal chromosomes, 2nd ed. San Diego: Academic Press, 1988.

Marlhens F, Chelly J, Kaplan JC, Lefrancois D, Harpey JP, Dutrillaux B. Familial deletion of Xp21.2 with glycerol kinase deficiency and congenital adrenal hypoplasia. Hum Genet 1987; 77:379-383.

Modi WS, Nash WG, Ferrari AC, O'Brien SJ. Cytogenetic methodologies for gene mapping and comparative analyses in mammalian cell culture systems. Gene Anal Techn 1987; 4:75-85.

Naylor SL, McGill JR, Zabel U. In situ hybridization of metaphase and prometaphase chromosomes. In: (M M Gottesman, ed.), Methods in enzymology. Vol 151: Molecular genetics of mammalian cells. San Diego: Academic Press, 1987.

Perry P, Wolff S. New Giemsa method for the differential staining of sister chromatids. Nature 1974; 251:156-158.

Popescu NC, Amsbaugh SC, Swan DC, DiPaolo JA. Induction of chromosome banding by trypsin/EDTA for gene mapping by in situ hybridization. Cytogenet Cell Genet 1985; 39:73-74.

Priest JH. Medical cytogenetics and cell culture. Philadelphia: Lea & Febiger, 1977.

Rigby PWJ, Dreckmann M, Rhodes C, Berg P. Labeling of deoxyribosenucleic acid to high specific activity in vitro by nick- translation with DNA polymerase I. J Mol Biol 1977; 113:237-251.

RØnne M. Chromosome preparation and high resolution banding techniques. A review. J Dairy Sci 1989; 72:1363-1377.

Rooney DE, Czepulkowski BH, eds. Human cytogenetics: a practical approach. Oxford, Washington, DC: IRL Press, 1986.

Schowalter, DB, Sommer SS. The generation of radiolabeled DNA and RNA probes with polymerase chain reaction. Analyt Biochem 1989; 177:90-94.

Seabright M. A rapid banding technique for human chromosomes. Lancet 1971; 2:971-972.

Wahl GM, Stern M, Stark GR. Efficient transfer of large DNA fragments from agarose gels to diazobenzyloxymethyl paper and rapid hybridization using dextran sulfate. Proc Natl Acad Sci USA 1979; 76:3683-3687.

Wolters WR, Chrisman CL, Libey GS. Lymphocyte culture for chromosomal analyses of channel catfish, *Ictalurus punctatus*. Copeia 1981: 1981:503-504.

7

Pulsed–Field Gel Electrophoresis and Its Application in the Physical Analysis of the Bovine Major Histocompatibility Complex

Albert Bensaid, John R. Young,* Anita Kaushal, and Alan J. Teale

International Laboratory for Research on Animal Diseases Nairobi, Kenya

INTRODUCTION

The power of resolution of in situ hybridization on chromosomes does not go below 1 million base pairs (1 Mbp) and DNA fragments of more than 20 kilobase pairs (kb) in length are not separated by conventional agarose gel electrophoresis. However, in 1984, Schwartz and Cantor (1) described an electrophoretic technique for the separation of yeast chromosomes. This technology, together with the availability of rare-cutter restriction enzymes, which cleave mammalian DNA into fragments ranging from 10 kb to 1 Mbp or more, means that it is now possible to examine, with reasonable definition, the physical relationships between components of mammalian multigene families and to establish linkage maps between genes separated by several hundred kilobases.

In this overview we will consider some aspects of the different techniques now available for the size separation of large DNA molecules,

**Present affiliation*: Agricultural and Food Research Council Institute for Animal Health, Compton, Newbury, Berkshire, England

emphasizing methods currently used in our laboratory. We will illustrate the use and value of the approach with examples taken from our studies of the bovine major histocompatibility complex (MHC).

PRINCIPLES AND TECHNICAL ASPECTS

Preparation and Digestion of High-Molecular-Weight DNA

This procedure may be used for the preparation of DNA from nucleated cells provided they are available as single-cell suspensions. Clumps must be avoided and cells counted accurately. Cell pellets are resuspended at 2×10^7 /ml in isotonic phosphate-buffered saline (PBS) and thoroughly mixed with an equal volume of 1.5% low-melt agarose in PBS at a temperature of 42°C. The mixture is then dispensed into 20 x 20 x 1 mm agarose block formers, on ice. It is important to use high-quality low-melt agarose, which can also be treated with DEAE-cellulose prior to use. This treatment eliminates most of the restriction enzyme inhibitors present in agarose. In our laboratory we routinely use GTG-grade Seakem (FMC Bioproducts). The thickness of agarose blocks should not exceed 2 mm in order to facilitate the diffusion of restriction enzymes in subsequent procedures and thus avoid partial digestion of DNA.

When the agarose has solidified, blocks are removed and gently placed in small beakers containing 10 ml of 10 mM Tris-HCl pH 8, 250 mM EDTA, 0.5% SDS, and 250 μg/ml of proteinase K. Beakers are then placed in a shaking water bath at 50°C for 36–48 h. Instead of SDS, some workers prefer to use sodium lauryl sarcosine at 1%. In our hands, both detergents work well. After protein digestion and membrane solubilization, the buffer is removed and replaced by 25 ml of 10 mM Tris-HCl pH 8/10 mM EDTA. Beakers are placed on shaking platforms for 1-2 h and a series of at least five washes is performed. Agarose blocks in 10 mM Tris-HCl pH 8/10 mM EDTA can be stored at 4°C for several months in Petri dishes.

This protocol yields high-molecular-weight DNA (more than 1 Mbp), which is trapped in the agarose pores. Unless degraded into small fragments (below 20 kb), the DNA will not diffuse and is suitable for digestion with restriction enzymes. Protease inhibitors such as PMSF or aprotinin can be added to inactivate remaining proteinase K.

For restriction enzyme digestion, agarose blocks are cut into small pieces of about 5 × 6 mm, which are distributed into 24-well plastic plates. The appropriate restriction enzyme (RE) buffer (about 1 ml) is added and the plate gently agitated for 1 h. The RE buffer is changed once and samples allowed to equilibrate for another hour. Small agarose slices are transferred to the bottom of 1.5-ml conical tubes in RE buffer (between 100 and 120 μl). The restriction enzyme is added and tubes are incubated at the appropriate temperature for 6–16 h.

Cutting and handling agarose slices is a tedious exercise, and most of the accidental degradations of DNA occur at this stage. It is therefore important to clean extensively with ethanol all the surfaces and spatulas that might be in contact with the agarose blocks. Since most of the enzymes used for pulsed-field techniques have been characterized by the suppliers, it is advisable to follow their instructions concerning the composition of the RE buffers and the optimal temperatures for DNA digestions.

Many parameters influence the cleavage of DNA trapped in agarose; quality of the agarose and thickness of the slice have been discussed. Mammalian genomes seem on average to contain about 40% G and C nucleotides, and the CG dinucleotide is fivefold rarer than expected from G+C content. Furthermore, long A+T stretches (with very few CG) are followed by islands rich in G+C in which the CG dinucleotide is found at the expected frequency. Therefore, the use of restriction enzymes recognizing G+C-rich sequences and/or CG dinucleotides on mammalian DNA will result in the generation of large DNA restriction fragments. In addition, most CG dinucleotides are methylated in mammals and therefore are generally resistant to enzyme cleavage. However, unmethylated CG dinucleotides reside on the G+C-rich islands, which are characteristically located 5' to genes (2–4). These considerations are important when interpreting patterns in which one enzyme gives rise to multiple bands. Indeed, two interpretations are possible in this circumstance: either the segment of DNA analyzed contains a multigene family, or the multiple bands observed are the result of partial digestion due to methylation. However, it is widely accepted that for a given cell type, CG sequences on given haplotypes are either completely methylated or unmethylated. Therefore, before reaching any conclusion on the organization of a certain region or, more important, on the multiplicity of genes, it is desirable to run digested DNA from different cell sources such as fibroblasts and peripheral blood lymphocytes (PBL).

The amount of restriction enzyme required to cleave high-molecular-weight DNA embedded in agarose varies with the enzyme used. In our hands, 40, 30, and 20 U of the enzymes *Mlu*I, *Sfi*I, and *Not*I, respectively, will achieve complete digestion of available sites (nonmethylated) in the system described in 6 h. Since the enzymes provided by the suppliers are of high quality, digestion can be carried out overnight. However, it is advisable to check samples for complete digestion by increasing both the quantities of enzyme and the periods of incubation. Figure 2 shows typical patterns obtained after digestion of bovine DNA with the restriction enzymes *Sfi*I, *Not*I, and *Mlu*I. Complete digestion can be monitored by the appearance of characteristic satellite bands.

Separation of High-Molecular-Weight DNA by Pulsed-Field Techniques

Although several apparatuses using different principles have been described, all achieve separation of large fragments of DNA on the basis of molecular weight. All the techniques may be termed pulsed-field.

Schwartz and Cantor (1) suggested that the separation of high-molecular-weight DNA molecules is the result of differing capacities to take angles. In contrast to the situation with conventional electrophoresis, where molecules are assumed to take a linear configuration for migration, changes in electric field direction in pulsed-field gel electrophoresis (PFGE) require molecules to alter their directions repeatedly. The hypothesis is that the higher-molecular-weight molecules do this more slowly than low-molecular-weight forms and thus their progress through the gel matrix is relatively delayed.

Pulsed-Field gel Gel Electrophoresis

This technique, described by Schwartz and Cantor (1), is based on the alternation of current applied to two sets of electrodes disposed as a square (Fig. 1a). Two electric fields are generated, one homogeneous and the other nonhomogeneous. The agarose gel is immersed in the

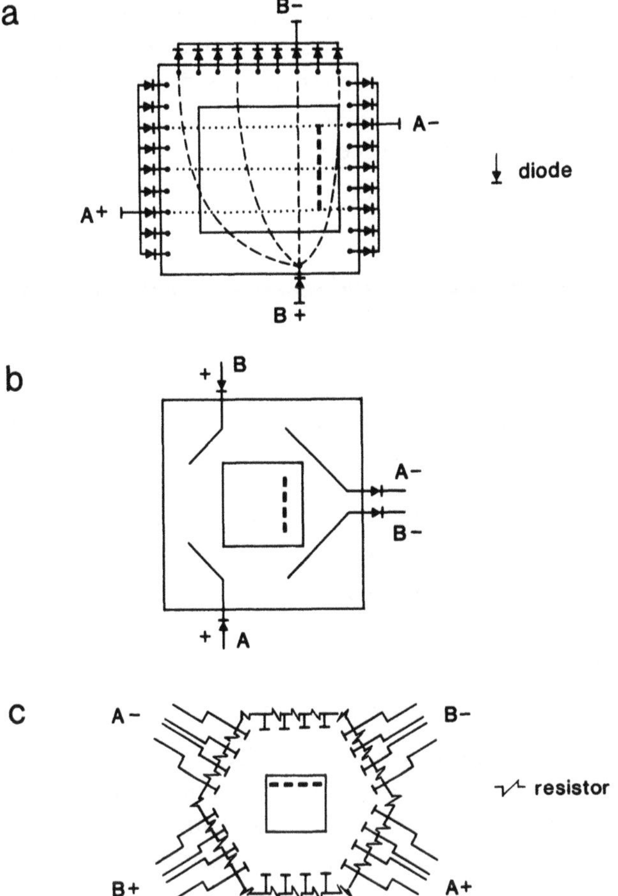

Figure 1 Diagram of the electrode arrays used for (a) PFGE, (b) OFAGE, and (c) CHEF.

electrophoretic buffer (generally TBE or TAE) and samples are loaded parallel to the electrodes providing the homogeneous field. The net effect of the two fields is that samples will migrate in an erratic manner, making lane-to-lane comparisons difficult. In addition, very-high-molecular-weight DNA is poorly resolved and appears as a bulb in which separation of bands is difficult to discern.

Despite the disadvantages, it is possible, using this system, to separate DNA fragments up to 2 Mbp. To obtain such separation, pulses of more than 100 s are required and electrophoresis is performed for more than 48 h. As high voltages (more than 300 v) are needed, systems for buffer recirculation and cooling are necessary. Also, larger tanks than for conventional electrophoresis are needed.

Orthogonal Field Alternation Gel Electrophoresis

In orthogonal field alternation gel electrophoresis, the same equipment is required as for PFGE. Here, the two sets of electrodes form an open angle (Fig. 1b), with the gel centralized (5). The current is alternatively switched between the two sets of electrodes, with each set providing a nonhomogeneous field. As a result of the action of the two nonhomogeneous fields, DNA fragments follow a curved course through the gel. The fanning observed at low molecular weights is due to a weakening field strength. Nonetheless, band patterns are sharper than in PFGE and direct comparison of DNA samples is facilitated. Similar conditions for running DNA are used in OFAGE and PFGE and similar resolutions are obtained. The bulb of high-molecular-weight DNA is slightly reduced in OFAGE.

Contour–Clamped Homogeneous Electric Field

Contour-clamped homogeneous electric field (CHEF), which was described by Chu et al. (6), is distinct from other systems in that it achieves separation of high-molecular-weight DNA by alternating two homogeneous fields. A homogeneous field is obtained by placing two infinitely long electrodes in parallel.

In practice, it is possible to produce a homogeneous electric field by using multiple electrodes arranged along a hexagonal contour in which two faces of the hexagon coincide with the positions of the infinite electrodes (Fig. 1c). The gel is placed in the middle of the electrode array in such a manner that the alternating electric fields have an angle of 120° relative to each other. The electrophoresis buffer is recirculated and cooled at 9°C. In the experiment described by Chu and colleagues, the pulse intervals were of 80 s and the field strength was 6 V/cm. The electrophoresis was performed 24 h. In these conditions, the largest chromosomes of *Saccharomyces cerevesiae* strain YNN295 could be resolved as sharp, defined bands, while in a parallel experiment performed by OFAGE, these chromosomes were present in a bulb at the top of the gel. Furthermore, in the CHEF system, samples migrated along straight paths except for a slight curvature of high molecular weights.

The CHEF system apparently overcomes many problems encountered with PFGE and OFAGE. However, the limitation of CHEF resides in the large dimensions of the electrophoresis tanks and the relative sophisticated electronics involved in building the electrode array. Effectively, each face of the hexagon comprises four electrodes connected by a series of resistors (470 ohms, 2 W). Although expensive, several CHEF systems are commercially available.

Field Inversion Gel Electrophoresis

In field inversion gel electrophoresis (FIGE), described by Carle and colleagues (7), the two electrodes are directly opposed and the electric field periodically inverted. Net migration is achieved by using a longer period for the pulse interval in one direction than in the reverse direction. If a single period ratio, for example 3 s forward and 1 s backward, is applied for 12–24 h, only low molecular weights ranging from 10 to 50 kb are properly resolved. If larger intervals are applied (e.g., 90 s forward and 30 s backward), DNA fragments of low molecular weight are compressed at the bottom of the gel but high-molecular-weight fragments ranging from 0.5 to 2 Mbp are separated. Therefore, to achieve a linear resolution of large and small DNA fragments, field interval ratios should be increased progressively following a gradient ramp.

In our laboratory, a program for controlling periodic field inversion was written for the Z80 processor (Torch, England) installed in an Acorn BBC model B microcomputer (Acorn Computers, England). The program was written in Pascal using a compiler (Prospero, England) which includes syntax for making calls to the Acorn operating system for timing and manipulation of the parallel port.

FIGE timing profiles are entered into the system using an interactive editing screen. Each profile can contain up to five sections. For the first section, initial and final forward and back field intervals can be specified directly or by entering one of them and a ratio. The period over which the change from the initial to final intervals is to occur is also entered. Subsequent sections start with the final intervals from the pre-

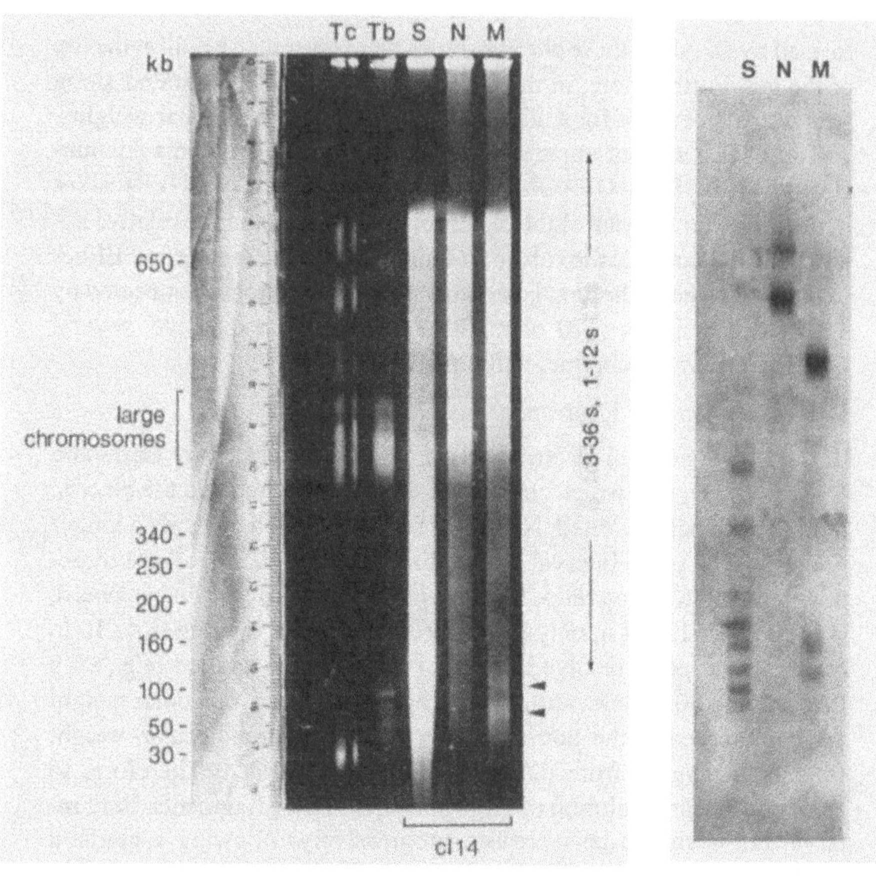

(a) (b)

Figure 2 Southern blot analysis of digested bovine DNA separated by FIGE. Bovine DNA, from a *Theileria parva*-infected lymphoblastoid cell line (B641.14), was prepared in low-melt agarose and digested with the enzymes *Sfi*I (S), *Not*I (N), and *Mlu*I (M). Restriction enzyme fragments were size-separated at 280 V for 28 h. At the beginning of the electrophoresis, the current was set forward for 3 s and then reversed for 1 s. It then followed a linear gradient ramp to reach 36 and 12 s in the forward and backward directions, respectively. (a) The gel was stained with ethidium bromide, and (b) DNA fragments were transferred into nylon filters, which were probed with an HLA- B7 class I gene. *Trypanosoma congolense* (Tc) and *Trypanosoma brucei* (Tb) chromosomes were run as molecular weight markers. Molecular weights (in kb) of *T. brucei* chromosomes are indicated at the left. Arrows indicate satellite DNA revealed by digestion with *Mlu*I.

ceding section, so only final values are specified. The program changes the pulse intervals in such a way that progression from initial to final intervals is a linear function of the elapsed time within each section. Specification of five separate intervals allows the overall profile of intervals to approximate many nonlinear functions of elapsed time. FIGE profiles are stored on diskettes for repeated use. During the running of a FIGE profile, the parameters and the current status are continually displayed. The current status is written to a diskette periodically in order to allow recovery from power failures.

Inversion of the field at the appropriate times is achieved by controlling a high-capacity relay with a smaller relay driven by a circuit with an independent power supply. The small relay is driven by the output of one bit of a parallel port (user port) of the microcomputer through an optical coupling circuit.

DNA fragments separated by FIGE have straight migrations, and therefore lane-to-lane comparisons are more accurate than with any other pulsed-field technique. Another advantage is that regular electrophoresis tanks are used and up to 18 samples per gel can be processed at a time. However, a good buffer cooling and recirculation system is required.

Many parameters can influence the migration of DNA in the gel. We have found that cooling the gel (10°C) from both above and below improved the sharpness of the bands. This prevents the establishment of a resistance gradient across the thickness of the gel. Reducing the thickness of the gel to 6–7 mm also helps in this respect. In addition, the gel should be covered by no more than 3 mm of buffer.

Another critical parameter is the concentration of DNA loaded in the low-melt agarose slice. The best results are obtained when no more than 2 μg of genomic DNA is contained in a 5 × 6 × 1 mm low-melt agarose slice. Increasing the concentration of DNA results in both poor resolution and delay in migration of the bands.

The greatest disadvantage of the FIGE system becomes apparent when short intervals are employed. In these circumstances very large DNA fragments can migrate faster than fragments of lower molecular weight. This inversion phenomenon is illustrated in Figure 2, in which the larger trypanosome chromosomes migrate ahead of the 650-kb chromosomes. When using more appropriate pulse intervals, such as those described in Figure 3, the larger trypanosome chromosomes are slower than the 650-kb chromosomes and are resolved in three bands.

Inversion of mammalian DNA fragments can be revealed either by running the digested samples with different pulse intervals or by performing double digestions. For example, when bovine DNA is cleaved with *Not*I and run with intervals, as described in Figure 2, a fragment of apparent molecular weight 500 kb is revealed by the human DRB probe. On the other hand, an *Sfi*I digestion results in the cleavage of the DRB region into two fragments of 1 Mbp and 260 kb. Thus, a double digestion with *Not*I and *Sfi*I should result in the disappearance of the 1 Mbp fragment. This is not the case, and upon double digestion with *Sfi*I and *Not*I, a band of 1 Mbp is still present. Thus, in these circumstances the *Not*I fragment containing the MHC class II DRB gene runs faster than the 1 Mbp *Sfi*I fragment, although its real size is more than 1 Mbp.

The reason for inversion is not fully understood. However, if the conformation of large DNA fragments in FIGE is compared to that of a spring, it is conceivable that under short pulse intervals very large DNA fragments do not have time to recover a compact conformation. Therefore, they will behave as if they were subjected to regular electrophoresis and will move with a stretched linear configuration. Other fragments of lower, molecular weight, in the same conditions, will adopt a compact configuration before changing to the opposite direction and therefore they will be retained longer in the agarose pores.

Two–Dimensional Gel Electrophoresis

Two-dimensional (2-D) electrophoresis of DNA is state of the art for physical gene mapping, but expensive. the procedure, which was described by Woolf et al., has been used to map the T-cell-receptor gamma gene family (8). High-molecular-weight DNA is prepared, digested, and run in very-high-purity low-melt agarose. The first dimension is performed by FIGE. Then, strips of agarose containing the size-separated DNA fragments are excised and equilibrated in a second RE buffer. The agarose strip is placed in a convenient container (a sealed 25-ml plastic pipet serves) and 200 U/ml of enzyme is added. The concentration of enzyme should be calculated to include the volume of the agarose gel. Following the second digestion, the agarose strip is sealed with agarose along a dialysis membrane onto which the DNA fragments

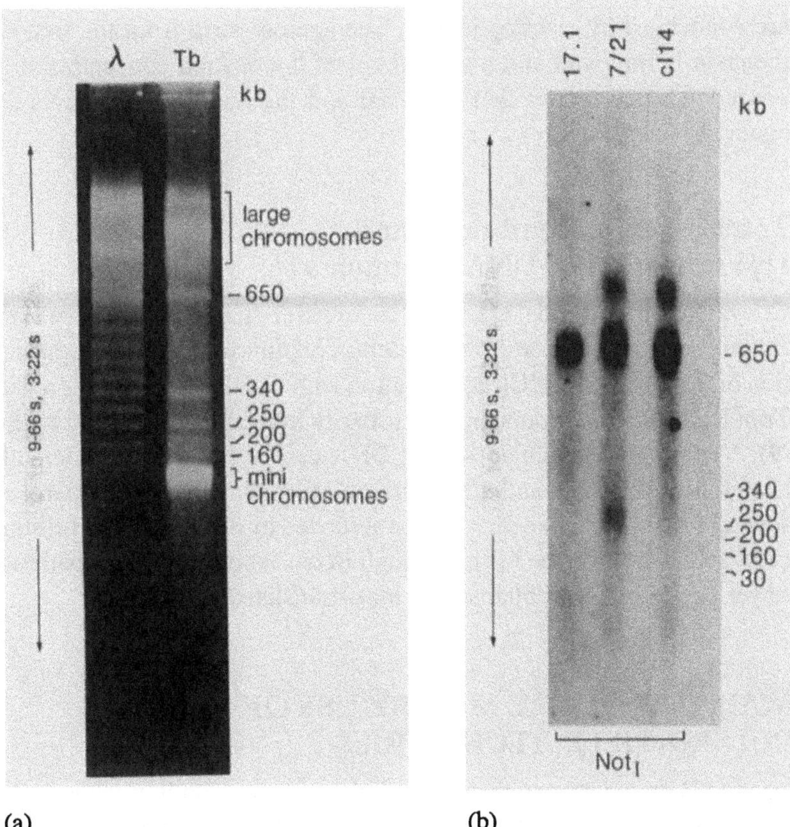

(a) (b)

Figure 3 Southern blot analysis of bovine DNA digested with *Not*I. Bovine DNA's from bovine cells (see Table 1 for phenotype) were prepared in low-melt agarose and digested with *Not*I. Electrophoresis was performed for 28 h at 280 V using pulse intervals of 9-66 s for the forward direction and 3-22 for the backward direction. (a) lambda DNA concatomers (monomer = 48 kb) and *T. brucei* chromosomes were run with the samples. Note that the large chromosomes are at the top of the gel. (b) DNA fragments were transferred onto nylon filters and probed with an HLA-B7 class I gene.

are condensed by electrophoresis. The agarose strip from the first dimension is removed and a second gel of the desired concentration is poured. The current is then reversed and the digested fragments are again size-separated.

Transfer onto Membranes and Hybridization of DNA Fragments

Conventional methods are used to transfer high-molecular-weight fragments from agarose gels to filters. This involves partial depurination of the DNA with HCl, denaturation with NaOH, neutralization with Tris-HCl pH 8, and transfer by capillarity in the presence of 20 × SSC (9). Very-high-molecular-weight DNA can be transferred efficiently to nylon membranes and after ultraviolet irradiation can be reprobed several times. Probe labeling can be achieved by either random priming (10) or nick translation (11), although in our laboratory we experience fewer background problems with nick-translated probes.

MAPPING OF THE MHC BY USE OF PULSED-FIELD TECHNIQUES

Snell (12) demonstrated that control of graft rejection in mice is under the control of multiple genes covering 2–3 cM of genome. An equivalent region has been described in all mammalian species examined, and a remarkable feature of the MHC, as it is now known, is the high degree of polymorphism displayed by some loci in most species (13).

The finding that MHC molecules are recognized by antigen-specific T lymphocytes raised on viral infection (14) added impetus to the exploration of this complex at the DNA level. It is now known that the complex encodes two types of such molecules, class I and class II, which restrict antigen recognition by specific cytotoxic T cells and T helper cells. Each subregion encoding class I and class II antigens is composed of several genes sharing common features. Thus the MHC is an archetypal example of a multigene family. With the introduction of pulsed-field techniques, physical maps of the MHC, measured in kilobases, can be obtained.

The Mouse H-2 Complex

Studies in recombinant mouse strains have revealed that the murine MHC comprises a class II region, H2-I, flanked by two class I regions, H2-K and H2-D (15). Genes encoding the C2 and C4 complement components were found to lie between the H2-I and H2-D regions; this region, which was also found to contain the gene coding for 21-hydroxylase, was called H2-S (16). Most of the murine MHC genes have now been cloned and ordered on the basis of overlaps in cosmid clones. The centromeric end of the mouse MHC was defined by a 600-kb cluster containing two and seven genes belonging to the H2-K and H2-I regions, respectively. The telomeric cluster, which comprised 13 class I and the tumor necrosis factor alpha and beta genes, was found to have a length of 500 kb. The H2-S region was placed between these two clusters, with a length of 300 kb. However, the orientation of the H2-S region and the physical distance separating this region from the telomeric and centromeric clusters remained unknown.

Linkage of the H2-I to the H2-S region was performed by pulsed-field techniques using probes defining the extremities of each region. Thus, the 21- hydroxylase gene was found 170 kb distal to the H2-I region. This gene was followed by the C4 complement gene and then the C2 complement gene, which defines the telomeric end of the H2-S region. The H2-D region could not be linked to the H2-S region; however, the distance separating these two regions was estimated to be at least 350 kb. Based on these results, the minimum length of the H2 complex was estimated at 2300 kb, which was in agreement with the genetic map derived from the study of recombinant mice (17).

The Human MHC

The strategy employed to study the human MHC (HLA) region was different from that used for the H-2 complex. The success of studies of the human MHC was mainly due to the development of outstanding serological reagents and techniques. Operationally monospecific antisera obtained from patients having rejected a graft or from multiparous women were used to screen human populations and large families.

At the beginning of the 1980s, it became apparent that the HLA region was organized differently from the mouse MHC. Three loci were found in the class I region, HLA-A, -B, AND -C. The frequency of recombination between the HLA-A and -B loci was 1%, while very little recombination was found between the HLA-B and -C loci. These studies indicated that the HLA-B and -C loci were tightly linked and the order from centromere to telomere was HLA-B, -C, and -A.

The C2 and C4 complement components as well as the 21-hydroxylase gene were placed centromeric to the HLA-B locus but telomeric to the HLA-DR class II region (18). Further investigation of HLA class II molecules was achieved by immunoprecipitations using mouse monoclonal antibodies and electrophoretic techniques. This led to identification of three sets of class II antigens, HLA-DR, -DQ, and -DP, each composed of one alpha chain and one beta chain. Here again, work performed at the DNA level resulted in dramatic progress in the understanding of the human MHC. The DR region was found to be composed of one alpha gene and three beta genes, while the DQ region comprises one alpha and one beta gene. In some individuals, the DQ region was found to be duplicated, as is the DP region. The fact that each class II probe displayed little or no cross-hybridization with other genes helped in performing further mapping studies (19).

Concerning the class I region, the generation of class I locus-specific probes was more difficult. Indeed, all the class I genes sequenced displayed between 75 and 95% identity. However, in most instances, the 3' untranslated (UT) regions of HLA class I genes and genomic DNA flanking the 5" termini of the genes were found to be operationally locus-specific (20,21). Although well characterized at the gene level, a physical linkage ordering and indication of the distance between genes were missing.

The first linkage map of the class II region was reported in 1986 by Hardy et al. (19). The technique used was OFAGE and the source of DNA was a lymphoblastoid cell line derived from a consanguineous homozygote at the MHC. Seven rare-cutter enzymes were used, individually and in combination, to digest DNA. The DP, DQ, and DR regions were found to occupy 100, 150, and 180 kb of DNA, respectively. The distances separating the DP from the DQ and the DQ from the DR region were 320 and 80 kb, respectively. In addition, two class II genes, DO and DZ, for which protein products are still unknown, were placed near the DQ and the DP region, respectively. Previous studies (22) indicated that the frequency of recombination between the DP and DQ regions was 1-3% which according to a genome average would be expected to correspond to a physical distance of 1000-3000 kb. The fact that pulsed-field experiments estimated this distance of 320 kb clearly demonstrated the presence of a recombination hot-spot between the DP and DQ regions. Conversely, no recombination was observed between the DR and the DQ regions, and this was in agreement with results obtained by pulsed-field experiments. Subsequently, two reports (23,24) confirmed this finding and linked the class II genes to the rest of the MHC.

Knowledge of the position and orientation of a gene cluster comprising the 21-hydroxylase and the C2 and C4 complement genes came with the analysis of overlapping cosmid clones. The 21-hydroxylase gene was placed telomeric to the DR alpha gene and at a distance of 300 kb. The gene was followed by the C4 and C2 complement genes, which are approximately 390 kb from the tumor necrosis factor (TNF) genes. A linkage between the TNF and the HLA-B class I gene was found; the genes being separated by 290 kb. The HLA-C class I gene was located centromeric to the HLA-B gene. Further studies in this region demonstrated that the HLA-B and C genes are separated by 130 kb (21). despite several attempts, including partial digestions with restriction enzymes, DNA fragments containing both HLA-B/C and HLA-A/E genes were not obtained. However, the HLA-E class I gene, for which a protein product is still unknown, was placed centromeric to the HLA-A gene and at a distance of 300-400 kb (24).

Although without an entire linkage map, the size of the human MHC was estimated at 3500 kb. A remarkable feature deduced from this genomic organization was that MHC genes are quite distant from one another. This observation suggests that many other genes may lie within the complex.

The Bovine MHC: Studies of Genetic Organization

The Class I Region

Studies of the bovine MHC (BoLA) system were first described in the 1970s (25). As with the HLA region, the initial approach involved generation of a battery of operationally monospecific alloantisera. International workshops have been held from time to time in order to assess and define progress (26). To date, 32 class I allospecificities have been identified at the international level and are referred to as BoLA-W1, -W2, and so forth.

It is still unclear whether these specificities are expressed on products of one or several loci. However, over the past several years increasing evidence has been obtained suggesting that multiple class I and II genes are expressed in cattle. At the protein level, by using a combination of mouse monoclonal antibodies and bovine alloantisera and monoclonal antibodies, it was possible to show that, on freshly collected peripheral blood lymphocytes, at least two different BoLA proteins could be encoded by a single haplotype (27,28). Further, when bovine lymphocytes are mitogen-activated, an additional population of class I-related molecules is expressed (29). Thus, the screening of BoLA class I diversity at the protein level strongly suggested that more than one functional gene encodes class I antigens in cattle.

Further evidence of multiple class I gene expression came with the study of gamma-irradiation and chemically induced deletion mutant lymphoblastoid cell lines. In these studies we were able to demonstrate the independent loss of expression of two sets of epitopes recognized by a large panel of monoclonal antibodies (MAb) specific for determinants encoded on each of two different bovine MHC haplotypes. These findings were supported by the transfection and expression in mouse L cells of the genomic DNA of an animal homozygous throughout the MHC. Two of the sets of epitopes (associated with a single haplotype), defined in the deletion mutant lines, were found to be independently expressed in the transfected L cells (30). Immunoprecipitations performed on these cells demonstrated that the BoLA class I molecules expressed had the same relative molecular mass as the molecules expressed on bovine PBL. The most probable interpretation of these results is that two different BoLA class I genes derived from a single haplotype were deleted in the mutant cell lines and the same genes were expressed in the mouse fibroblasts.

As in other species, a great advance in the understanding of the MHC was achieved with the cloning of bovine class I genes. In our laboratory, a cDNA library was made from DNA purified from PBL of an MHC homozygous animal. The animal used as the PBL donor was also the donor of the genomic DNA used in the transfection experiments already referred to. The first screen of this library was performed with an entire bovine cDNA derived from a different library (provided by Dr. Pamela Brown, AFRC Institute for Animal Physiology and Genetics Research, Edinburgh). Several class I clones were identified, and two of them (2.1 and 5.1) were shown to be distinct from each other on the basis of sequence and restriction mapping. The two 3'UT regions displayed 80% homology. As will be described, they were further used for mapping purposes.

In other species (human and mouse), it has been shown that the 3'UT regions of class I MHC genes are not often truncated and therefore can only generate variability at the messenger level by alternative splicing with the last coding exon of the molecule. The bovine 2.1 and 5.1 3'UT regions were of similar size (250 bp). Several stretches of five and eight nucleotide substitutions were found, and these were scattered rather than clustered. These findings indicated that the two cDNAs are certainly the products of two distinct class I genes and not the result of alternative splicing occurring in a single messenger RNA.

Studies were undertaken of a cloned lymphoblastoid cell line B641.14 (derived from an MHC heterozygous Boran *Bos indicus* animal) and of gamma-irradiation and chemically induced MHC deletion mutant subclones. The parent line possessed BoLA haplotypes carrying W10 and KN18 class I genes at the A loci. On the basis of studies with class II-specific alloreactive T-cell clones and isoelectric focusing of class II molecules (performed for us by Dr. I. Joosten, University of Utrecht), these haplotypes are clearly distinct in the class II region. With respect to the W10-encoding haplotypes, this also appears, on the basis of deletion mutant and transfection studies already described, to encode a second class I gene, KN104, the product of which is recognized by MAb and alloantisera.

Following digestion of the DNA of the animal from which the parent line B641.14 was derived, with the restriction enzyme *Not*I, two fragments of 650 and 1000 kb could be revealed by an HLA-B7 class I probe. In contrast, the irradiation-mutated clone 17.1 (loss of class I expression due to deletion on the W10 haplotype) had lost a fragment of 1000 kb and clone 7/21 (loss of class I expression due to a deletion on the KN18 haplotype) had apparently gained a fragment of 250 kb (Fig.3). The possibility that in the case of 7/21, gamma irradiation had created DNA was excluded when using other restriction enzymes (Fig. 5).

However, from this analysis, it was clear that the loss of allospecificities in the mutated clones is accompanied by deletion of genetic material or important modifications of the class I region. Proof that the 650- kb *Not*I fragment contained the genes coding for the w10 and KN104 allospecificities was provided by using class I cDNA 3'UT region probes on *Not*I-digested genomic DNA of several animals, including one homozygous at the MHC, and clone 17.1 (Fig. 4). Upon hybridization with an entire class I cDNA, two fragments of 650 and 1000 kb were revealed in all DNA tested except that of 17.1, which displayed only a 650-kb fragment. Thus it appeared that the mutation involved an extensive deletion in clone 17.1 resulting in the loss of all class I-expressed genes present on the w10,KN104 haplotype. Assuming that the mutation in 17.1 affected only one haplotype, it is likely that the one encoding KN18 is composed of one or more 650 and 1000-kb *Not*I fragments. Since a 250-kb *Not*I fragment appears in clone 7/21 (KN18 loss mutant), the gamma irradiation appears to have induced a partial deletion of the class I genes of the KN18 haplotype which is less extensive than in 17.1

Southern blots performed following digestion with *Sfi*I were very informative (Fig. 5). The parental clone, B641.14, revealed eight *Sfi*I fragments spanning from 370 to 45 kb (Table 1). On the other hand, clones 17.1 and 7/21 displayed four and six *Sfi*I fragments, respectively. Remarkably, the superimposition of the patterns provided by the mutated clones produced the pattern of the parental clone. The fact that clone 7/21 had fewer fragments than B641.14 demonstrated that deletions of DNA occurred in the haplotype coding for the BoLA KN18 gene. Moreover, because there was no simultaneous disappearance of *Sfi*I fragments having the same size on clones 17.1 and 7/21,it was deduced that the gamma-irradiation affected only one haplotype in each clone. Two *Sfi*I fragments of 100 and 60 kb were present in all three lines. This supported the view that one part of the MHC containing class I sequences was not affected by the gamma-irradiation in clone 7/21. Whether this region is able to code for conventional or class I-like genes remains unknown.

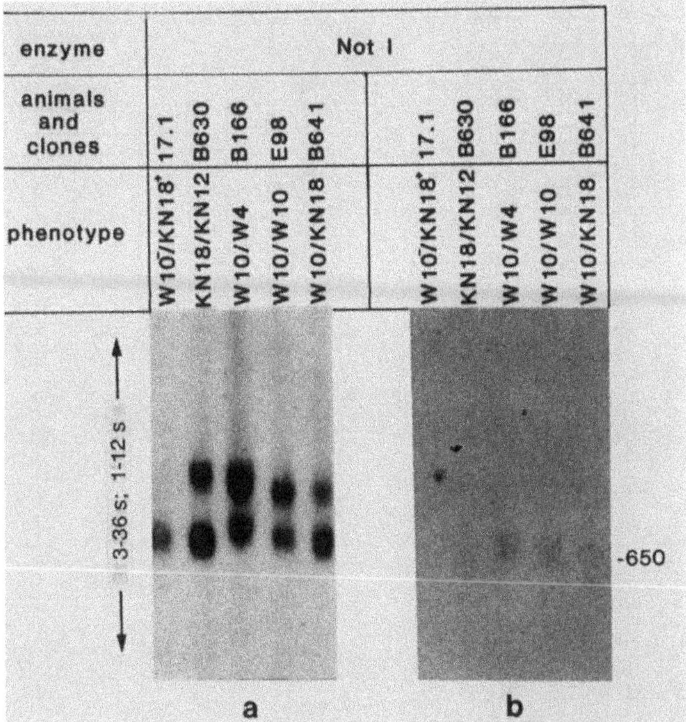

Figure 4 Comparison of *Not*I fragment derived from homozygous, heterozygous animals and a mutated clone line. DNA was prepared and electrophoresed as described in Figure 2. The probes were (a) a class I HLA-B7 gene and (b) the 3'UT region of the cDNA clone 2.1. At the right, the 650-kb *T. brucei* chromosome is indicated as molecular weight marker. Class I phenotypes, type of cells used, and restriction enzyme are indicated at the top of the autoradiograph. As shown in Figure 6, hybridizations of the 3'UT probes are more effective in the w10, KN104 haplotype than in the KN18 haplotype.

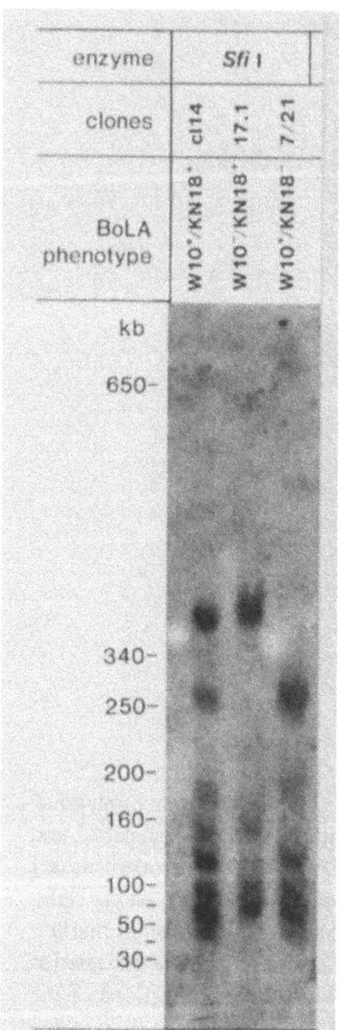

Figure 5 Southern blot analysis of bovine DNA from deletion mutant clones and the parental cloned line. DNA preparation and electrophoresis were performed as described in Figure 2. Samples obtained from the parental clone, B641.14, and the class I deletion mutant clones 17.1 and 7/21 were run side by side. Blotted DNA was probed with a class I HLA-B7 gene. *T. brucei* chromosomes are indicated at the left and size shown in kb.

Table 1 Size (in kb) of Restriction Fragments Generated with DNA of an MHC Homozygous Animal, a Heterozygous Lymphoblastoid Cell Line, and Two Derived Deletion Mutant Clones, Hydridizing with Class I Probes Following Digestion with Three Different Enzymes

E98 (w10, KN104 homozygous)	B641.14 (w10, KN104/ Kn18)	17.1 (w10, KN1O4 deleted)	7/21 (KN18 deleted)
Sfi I			
—	370	370	—
250	250	—	250
180	180	—	180
—	160	160	—
120	120	—	120
100	100	100	100
60	60	60	60
45	45	—	45
Mlu I			
450[a]	500	500	500
—	150	150	—
140	140	—	140
120	120	—	120
50[a]			
Not I			
1000	1000	—	1000
650	650	650	650
			350

— indicates that fragments present in B641.14 are not found in the DNA of animal E98 or the deletion mutant clones.

[a]An extra MluI site not found in B641.14. Whether this site has been created or unmethylated remains unknown.

Underlined are the fragments hybridizing with the 3'UT regions of bovine cDNA clones 2.1 and 5.1.

A linkage between the class I region coding for the allospecificities and the additional class I gene region was impossible to demonstrate. Further, even though fragments deleted in 17.1 and 7/21 could be assigned to specific haplotypes, because multiple fragments were involved it was impossible to connect individual ones with the w10,KN104 specificities, on the other hand, and with the KN18 specificity, on the other, Therefore, a search for restriction fragment length polymorphism in our cattle population was undertaken with *Sfi*I.

Analysis of 22 animals, together expressing 14 BoLA class I allospecificities, revealed that the 250-kb and the 370-kb *Sfi*I fragments correlated with expression of the w10 and KN18 allospecificities, respectively (Bensaid et al., in preparation).

Demonstration that the 250-kb and the 370-kb *Sfi*I fragments are able to be translated as messenger RNAs was provided by applying the class I cDNA 3'UT probes previously described (Fig. 6). Both probes, 5.1 and 2.1, hybridized preferentially to a 250-kb *Sfi*I fragment on B641.14 DNA and DNA obtained from peripheral blood lymphocytes of the w10,KN104 homozygous animal E98. However, the 3'UT probe 2.1 hybridized also with a 370-kb *Sfi*I fragment on B641.14. These results showed that, to some extent, 3'ut probes in cattle are locus-specific.

Double-digestion experiments using *Sfi*I, *Not*I, *Mlu*I, and *Cla*I in all pairwise combinations were performed in order to map the BoLA class I region (Table 2). It appeared that the 250-kb *Sfi*I fragment did not contain any *Not*I or *Mlu*I sites, indicating that this *Sfi*I fragment is entirely contained in the 650-kb *Not*I and the 500-kb *Mlu*I fragments, respectively. However, double digestions performed with *Sfi*I and *Cla*I resulted in detection of a 200-kb fragment hybridizing with both probes, 5.1 and 2.1. Thus, if the two cDNAs are the product of two different class I genes, pulsed-field analysis demonstrates that these expressed genes lie within a DNA fragment of 200-kb. In the same haplotype (w10/KN104) only one *Sfi*I fragment of 180 kb was lost on digestion with *Not*I. This generated two fragments of 140 and 40 kb containing class I region sequences. Therefore, one of the extremities of the 1000-kb or the 650- kb *Not*I fragments contains class I genes. The fact that upon double digestion two fragments hybridized to the class I gene probe indicated that the 180-kb *Sfi*I fragment might be a good candidate

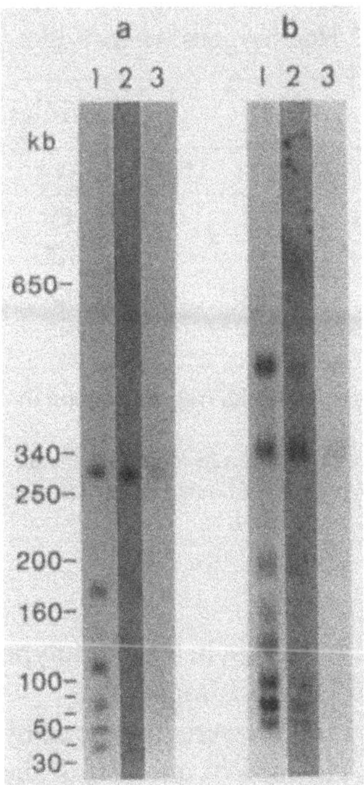

Figure 6 Definition of class I locus-specific probes in cattle. DNA from the homozygote B641 (b) was digested with *Sfi*I. Electrophoresis was performed for 28 h at 280 V using pulse intervals of 3-24 s and 1-8 s in the forward and backward directions, respectively. DNA fragments were blotted onto nylon filters and consecutively probed with an entire bovine class I cDNA (1), the 3'UT region of the class I cDNA 2.1 (2), and the 3'UT region of the class I cDNA 5.1 (3). *T. brucei* chromosomes are indicated at the left.

to link the 1000-kb and the 650-kb *Not*I fragments. However, in the absence of locus-specific probes for these class I genes, this remains speculative.

On the second haplotype of B641, which codes for the KN18 allospecificity, the 370-kb *Sfi*I fragment displayed two internal sites for *Not*I and *Mlu*I (Table 1). In both instances, double-digestion experiments revealed the presence of an extra fragment at 40 kb, as assessed

Table 2 Size (in kb) of Restriction Enzyme Fragments Obtained upon Single and Double Digestion of DNA of the MHC Homozygous Animal E98

SfiI	SfiI/NotI	SfiI/MluI
250	250	250
180	140 + 40[a]	45[b]
120	120	120
100	100	45[b]
45	45	45

[a]The 180-kb SfiI fragment has been cleaved into 140- and 40-kb fragments upon digestion with NotI.
[b]The 180- and 100-kb SfiI fragments have been cleaved by MluI into 45-kb fragments. Underlined are the fragments hybridizing with 3'UT regions of bovine cDNA clones 2.1 and 5.1.

by analysis of patterns obtained with clone 17.1. Thus, in this haplotype also, the BoLA class I region spans beyond the *Not*I fragments.

Further attempts to map the BoLA class I region were hampered by lack of probes specific for each subregion. The map of one part of the class I region presented in Figure 7 should be considered tentative. Nonetheless, from the information reported in this study, it is possible to sketch the organization of the BoLA class I region. The two *Not*I fragments together gives a length for the BoLA class I region of 1650 kb, which is also the estimated size of the human class I MHC. The fact that several *Sfi*I fragments can be cleaved by *Not*I generating two fragments with class I sequences strongly suggests that the class I complex spans 1650 kb. Another estimate, which should be considered the minimum size of the class I region, is achieved by adding all *Sfi*I fragments. By doing so, the size of the class I region would be 770 kb for the E98 haplotype coding for the w10 and KN104 allospecificities. It is unlikely that all the *Sfi*I sites are contiguous, as studies performed in the human system tend to demonstrate. The same exercise performed with *Mlu*I gives an estimate of 760 kb.

*Sfi*I cleaves the class I complex into six fragments with E98 DNA. Assuming that one BoLA class I gene measures 5 kb, and that the probability of there being an *Sfi*I site in any given 5-kb piece of DNA is low,

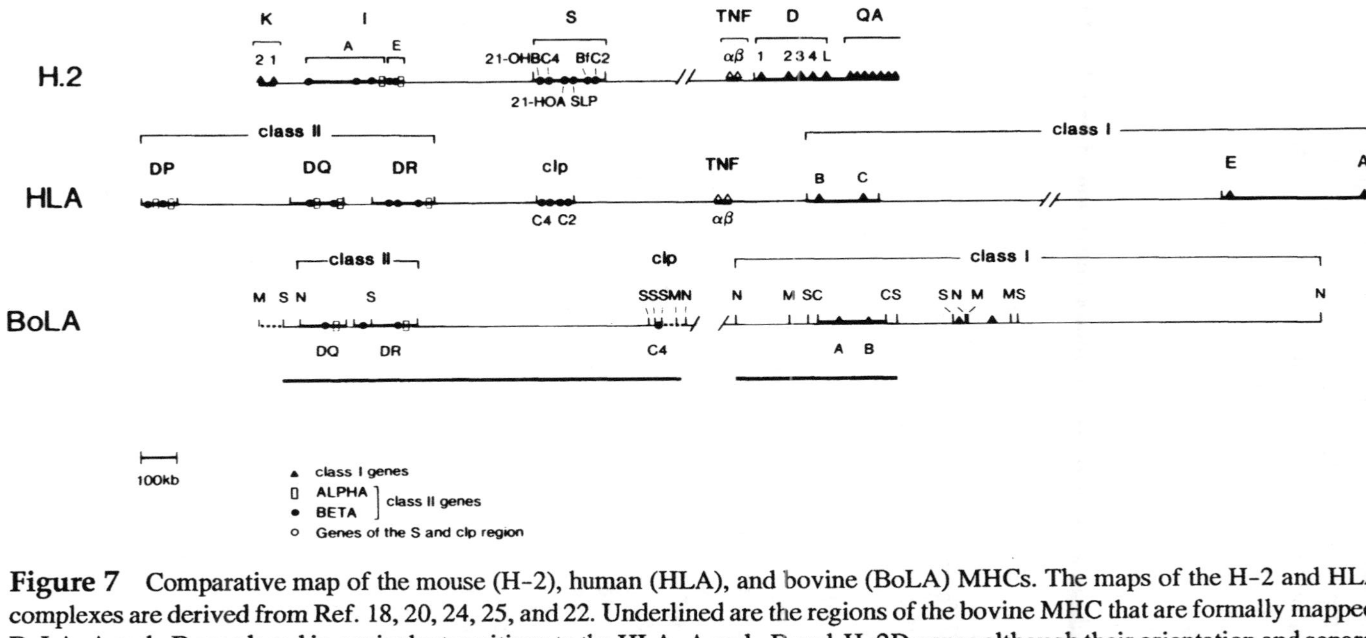

Figure 7 Comparative map of the mouse (H-2), human (HLA), and bovine (BoLA) MHCs. The maps of the H-2 and HLA complexes are derived from Ref. 18, 20, 24, 25, and 22. Underlined are the regions of the bovine MHC that are formally mapped. BoLA-A and -B are placed in equivalent positions to the HLA-A and -B and H-2D genes although their orientation and separation distance are unknown. Fragments containing class I genes with unknown products are not designated with letters. The restriction enzyme map is indicated only for the bovine MHC; S = *Sfi*I, N = *Not*I, M = *Mlu*I. Abbreviations for genes are; 21-OH = 21-hydroxylase, C4 = C4 complement component.

it is probable that every *Sfi*I fragment contains at least one class I gene. this is exemplified by the finding that two distinct 3'UT regions are transcribed by the same *Sfi*I fragment of 250 kb. Thus two genes are encountered in this fragment, which have to be added to the genes contained in the five remaining *Sfi*I fragments, giving a minimum estimate of seven genes located in one haplotype.

An important question remaining concerns the function of these additional genes. Our previous work on the expression of class I antigens revealed that upon T-cell growth factor activation of peripheral blood lymphocytes, additional class I-like molecules were expressed at the cell surface. The molecules were associated with beta 2-micro-globulin, shared some monomorphic epitopes with conventional class I antigens, but differed in having a smaller protein backbone (29). A differential screening of cDNA library obtained from activated T cells should provide cDNA clones coding for these proteins.

The Class II Region

Studies using conventional molecular biological techniques have already generated a substantial amount of information concerning the bovine class II region, BoLA-D. By analogy with the human class II region, two BoLA-D subregions, DR and DQ, have been identified. The human probes that were used were generally locus-specific in cattle, although some degree of cross-reaction was noticed between the DR beta (DRB) probe and the bovine DQ beta (DQB) genes. The BoLA-DR region was found to comprise one DR alpha (DRA) gene and at least two DRB genes, while in most instances the DQ region was composed of one alpha and one beta gene. However, in some animals, it appeared that the DQA-B tandem was duplicated. From these RFLP studies it was deduced that the BoLA-D and HLA-D regions were very similar and should therefore have a similar organization (31).

This conclusion was supported by studies in our laboratory using pulsed-field techniques. The physical map of the bovine class II region was deduced from the DNA analysis of 22 animals representing 14 different haplotypes (as assessed by class I serological typing). Eight of the animals were homozygous throughout the MHC. Upon digestion of DNA with *Not*I, *Sfi*I, and *Mlu*I and consecutive hybridization with the human DRA, B and DQA and B probes, a linkage between bovine DR

Table 3 Size (in kb) of SfiI Fragments Generated by Digestion of DNA of Four Cattle with SfiI

Animal/class I	DRA/B	DRB	DRB-DQA/B	DQA/B
B630(KN18/KN12)	340		260 200	
B641(W10, KN104/KN18)	1000 340		260	
E98(W10, KN104)[a]	1000		260	
E37 (KN12)[a]	1000	50		170

[a]Animals E98 and E37 are homozygous at the MHC.

and DQ was evident. Indeed, all four probes hybridized to the same fragments generated by *Not*I and *Mlu*I (Table 3 and Fig. 7). Patterns obtained with *Sfi*I were more complex in that they revealed several polymorphic fragments ranging from 1000 kb to 50 kb. In the population, the DRA gene was in a 1000-kb or 340-kb *Sfi*I fragment, and occasionally in a 500-kb *Sfi*I fragment. These fragments hybridized also with the DRB probe, proving that DRA and B genes are not separated by more than 340 kb. In addition, three more polymorphic bands of 260, 200, and 50 kb containing DRB but not DRA genes were observed in the population. Probing with the DQA and DQB human cDNAs resulted in identification of three *Sfi*I fragments of 260, 200, and 170 kb. The two largest fragments were shown to contain DRB sequences. Thus, DRB and DQA and B genes are linked and occupy a maximum of 200 kb of DNA.

These findings were supported by performing double digestion with *Sfi*I and *Not*I on DNA of the homozygous animal E98. The *Sfi*I/*Not*I fragment was 200 kb and contained DRB and DQA and B sequences. In some homozygous animals (carrying different BoLA haplotypes), in addition to a 1000-kb *Sfi*I fragment that contains both DRA and B genes, a 50-kb *Sfi*I fragment hybridized only with the DRB probe, while the DQ genes were found in a 170-kb *Sfi*I fragment. These results show that the second set of DRB genes does not span more than 50 kb and that the DQA and B genes do not occupy more than 170 kb.

In all animals tested, using several rare-cutter enzymes, it was impossible to separate DQA from DQB genes, indicating a very close linkage between them. Recently it was shown, by conventional analysis of a lambda genomic library, that 15 kb of DNA separates the DQA gene from the DQB gene. The fact that a lambda clone containing DRA and DRB genes was not found, strongly suggested that the genes are at least 20 kb apart (32).

Attempts to define a DP-like region in cattle have not been successful. Indeed, the DPB human probe used in these studies hybridized with fragments already defines by the DRB probe. Interestingly, small *Sfi*I fragments (50 kb) containing DRB genes did not hybridize with the human DPB cDNA. This may suggest that DRB genes are heterogeneous within the cattle population. Recent studies have pinpointed that human DRB cDNAs weakly hybridized with restriction enzyme DNA fragments that defined a new polymorphism. Population and family studies indicated that this region, known as DY, had a frequency of recombination of 3% with the DR-DQ region (31). Further, a clone from a lambda genomic library was found to contain a DRB-like gene, as assessed by partial sequencing (32). It is possible, therefore, that the bovine DY and the human DP regions are equivalent but have considerable diverged during evolution. However, a formal linkage between the DR-DQ and the DY region remains to be demonstrated. Here again, pulsed-field techniques may prove to be useful. So far, the DRB probe has been shown to hybridize weakly with *Not*I and *Sfi*I fragments not revealed by the DRA probe. In contrast to the real DRB genes, the light fragments were not polymorphic when DNA was digested with *Sfi*I. When available, the DY probes should demonstrate whether or not there is a DP-like region in the bovine MHC.

Linkage of the Class II MHC Region and the C4 Complement Gene

The linkage between complement and class II genes in cattle has been investigated by conventional RFLP studies. It appeared that the C4 complement gene was linked with the class I and class II regions and therefore that the C4 gene is probably positioned between them (31).

Using a human C4 complement probe on bovine DNA digested with rare–cutter enzymes, we have been able to establish a physical linkage between the bovine C4 gene and the class II DR and DQ genes. Indeed, a *Not*I fragment of more than 1000 kb known to contain all DR and DQ genes hybridized with the C4 probe. However, in some haplotypes studied, a different *Not*I fragment of 150 kb also displayed the presence of a C4 gene. Interestingly, and in contrast with the class I and class II situation, the enzyme *Sfi*I did not reveal polymorphism. In all animals examined, a single *Sfi*I fragment of 20 kb hybridized to the C4 probe but not to DR, DQ, and class I genes. Therefore, complement genes are linked to class II genes but the distance separating them is substantial. This distance can be estimated to be a minimum of 650 kb, assuming that the DRA and B genes do not span more than 350 kb and that *Sfi*I fragments containing C4 and class II genes, respectively, are contiguous.

We have not been able to establish a physical linkage between complement and class I genes. This result can be interpreted in two ways. The first possibility is that class I and complement genes are very distant from each other. The second is that an unusual number of *Not*I, *Mlu*I, and *Sfi*I sites are present between the class I region and the C4 complement gene.

Comparative Map of the H–2, HLA, and BoLA Complexes

It is now possible to construct, on a megabase scale, a comparative map of the mouse, human, and bovine MHC regions, as shown in Figure 7. The class II DR– DQ, complement, and class I regions of humans and cattle can be aligned with the mouse H–2 I/E, H–2 S, and H–2 D/Qa regions, respectively. Of particular interest, is that the distance separating the DRA gene from the C4 gene in cattle is much greater than in mouse and human. As a consequence, this portion of the bovine genome may contain several functional genes still to be discovered in cattle. Such putative genes may account for some of the associations that have been reported between MHC type and disease resistance/susceptibility and performance in cattle.

Further differences between the species are apparent on the centromeric side of the MHC regions. First, the mouse complex, unlike those of human and cattle, includes class I genes in this area. The HLA-DP region has no apparent homolog in mouse or cattle. By revealing such differences between species, pulsed-field techniques, through the provision of maps of multigene families spanning hundreds of kb of DNA, is providing information on basic evolutionary mechanisms.

ACKNOWLEDGMENTS

We thank Dr. S. Kemp and Mr. J. Mwakaya for assistance with MHC phenotyping, Mr. A. Lobo and Mr. J. Magondu for assistance with the design and construction of the FIGE apparatus and Dr. R. L. Spooner of the AFRC, Institute for Animal Physiology and Genetics Research, Edinburgh, for provision of BoLA typing alloantisera. We are also grateful to Dr. F. Lemmonier of CIML, Marseille, for the HLA-B7 gene probe and to Dr. L. Andersson of the Swedish Agricultural University, Uppsala, for the class II probes.

REFERENCES

1. Schwartz DC, Cantor CR. Separation of yeast chromosome-sized DNAs by pulsed field gel electrophoresis. Cell 1984; 37:67-75.
2. McClelland M, Ivarie R. Asymmetrical distribution of CpG in an "average" mammalian gene. Nucl Acids Res 1982; 10:7865-7877.
3. Brown WRA, Bird AP. Long range restriction site mapping of mammalian genomic DNA. Nature 1986; 323:477-481.
4. Lindsay S, Bird AP. Use of restriction enzymes to detect potential gene sequences in mammalian DNA. Nature 1987; 327:336-338.
5. Carle GF, Olson MV. Separation of chromosomal DNA molecules from yeast by orthogonal field alternation gel electrophoresis. Nucl Acids Res 1984; 12:5647-5664.
6. Chu G, Vollrath D, Davis RW. Separation of large DNA molecules by clamped homogeneous electric fields. Science 1986; 234;1582-1585.
7. Carle GF, Frank M, Olson MV. Electrophoretic separation of large DNA molecules by periodic inversion of the electric field. Science 1986; 232:65-68.
8. Woolf T, Lai E, Kronenberg N, Hood L. Mapping genomic organization by field inversion and two dimensional gel electrophoresis: application to

the murine T-cell receptor gamma gene family. Nucl Acids Res 1988; 16:3863- 3875.

9. Maniatis T, Fritsch E F, Sambrook J. Molecular cloning: A laboratory manual. New York: Cold Spring Harbor Laboratory Press, 1982.
10. Feinberg AP, Vogelstein BA. Technique for radiolabeling DNA restriction endonuclease fragments to high specific activity. Analyt Biochem 1983; 132:6-13.
11. Rigby PWJ, Diekmann M, Rhodes C, Berg P. Labeling deoxyribonucleic acid to high specific activity in vitro by nick translation with DNA polymerase I. J Mol Biol 1977; 113:237-251.
12. Snell GD. Methods of the study of histocompatibility genes. J Genet 1948; 49:87-99.
13. Klein J. Generation of diversity at MHC loci: Implication for T cell receptor repertoire. In: Fougereau M, Dausset J, eds. Immunology 80. London; Academic Press, 1980:239-251.
14. Zinkernagel RM, Doherty PC. Restriction of in vitro T cell-mediated cytoxicity in lymphocytic choriomeningitis within a syngeneic or semiallogeneic system. Nature 1974; 248:701-702.
15. Klein J. Biology of the mouse histocompatibility-2 complex. New York: Springer Verlag, 1975.
16. Schreffler DC, Owen RD. A serologically detected variant in mouse serum: inheritance and association with the histocompatibility-2 locus. Genetics 1963; 48:9-35.
17. Muller V, Stephen D, Philippsen P, Steinmetz M. Orientation and molecular map position of complement genes in the mouse MHC. EMBO J 1987; 6:369-373.
18. Dausset J, Pla M, eds. HLA the human major histocompatibility complex. Paris: Flammarion, 1985.
19. Hardy DA, Bell JI, Long EO, Lindstein T, McDevitt HO. Mapping of the class II region of the human major hsitocompatibility complex by pulsed field gel electrophoresis Nature 1986; 323:453-455.
20. Strachan T, Dodge AB, Smollie D, Dyer PA, Sodoyer R, Jordan BR, Harris R. An HLA-C specific DNA probe. Immunogenetics 1986; 23:115-120.
21. Pontaroti P, Chimini G, Nguyen C, Boretto J, Jordan BR. CpG islands and HTF islands in the HLA class I region: investigation of the methylation status of class I genes leads to precise physical mapping of the HLA-B and C genes. Nucl Acids Res 1988; 16:6767-6777.

22. Termitjelen A, Meera Khan P, Shaw S, Van Rood JJ. Mapping SB in relation to HLA and GLO1 using cells from first-cousin marriage offspring. Immunogenetics 1983; 18:503-512.
23. Dunham I, Sargent CA, Trowsdale J, Campbell RD. Molecular mapping of the human major histocompatibility complex by pulsed-field electrophoresis. Proc Natl Acad Sci USA 1987; 72:37-41.
24. Carroll MC, Katzman P, Alicot EM, Koller BH, Geraghty DE, Orr HT, Strominger JL, Spies T. Linkage map of the human major histocompatibility complex including the tumor necrosis factor genes. Proc Natl Acad Sci USA 1987; 84:8535-8539.
25. Spooner RL, Oliver RA, Sales DI, et al. Analysis of alloantisera against bovine lymphocytes. Joint report of the 1st International Bovine Lymphocyte Antigen (BoLA) Workshop. Anim Blood Groups Biochem Genet 1979; 15:63-86.
26. Bull RW, Lewis HA, Wu MC, et al. Joint Report of the Third International Bovine Lymphocyte Antigen Workshop, Helsinki, Finland, July 27, 1987. Animal Genetics 1989; 20:109-132.
27. Bensaid A, Naessens J, Kemp SJ, Black SJ, Shapiro SZ, Teale AJ. An immunochemical analysis of class I (BoLA) molecules on the surface of bovine cells. Immunogenetics 1988; 27:139-144.
28. Kemp SJ, Tucker EM, Teale AJ. A bovine monoclonal antibody detecting a class I BoLA antigen. Animal Genetics 1990; 21:153-160.
29. Bensaid A, Kaushal A, MacHugh ND, Shapiro SZ, Teale AJ. Biochemical characterization of activation-associated bovine class I major histocompatibility complex antigens. Animal Genetics 1989; 20:241-225.
30. Toye P, MacHugh ND, Bensaid A, Alberti S, Teale AJ, Morrison WI. Transfection into mouse L cells of genes encoding two serologically and functionally distinct bovine class I MHC molecules from a MHC-homozygous animal: evidence for a second class I locus in cattle. Immunology 1990; 70:20-26.
31. Anderson L, Lunden A, Sigurdardottir S, Davies CJ, Rask L. Linkage relationships in the bovine MHC region. High recombination frequency between class II subregions. Immunogenetics 1988; 27:373-280.
32. Groenen MAM, Van Der Poel JJ, Dijihof RJM, Giphart MJ. Cloning of the bovine major histocompatibility complex class II genes. Animal Genetics 1989; 20:267-278.

8

Chromosomal Assignment of Marker Loci to Monitor Marker Spacing

Ruedi Fries, Sabina Solinas, and Asoka Gunawardana

Federal Institute of Technology, ETH-Zentrum
Zurich, Switzerland

INTRODUCTION

The information from gene maps consists of two elements: information about the physical and the genetic map position. The physical map is established by assigning genes to particular chromosomes or regions of a chromosome. Genetic maps are constructed by studying the meiotic linkage relationships of gene loci (linkage analysis).

The genetic map will be the basis for establishing the linkage relationship of marker loci with genes of interest for animal breeding. The genomes of domestic species may encompass about 3000 cM (1). For genetic mapping one would want to have all points of the genome within not much more than 20 cM of a mapped marker locus, in order to enable linkage relationships to be established with a reasonable number of animals. Thus, on an average bovine chromosome of 100 cM, one would need three well-spaced markers per chromosome—two about 20 cM from the centromere and the telomere, respectively, and one in the center; in swine, one would need four markers on an average chromosome of about 160 cM. That is, about 90 marker loci, well spaced along the chromosomes, would initially be sufficient for complete coverage of the 30 chromosome pairs of cattle or less than 80 would be sufficient to cover the 19 porcine chromosome pairs. However, since marker spacing cannot be directed, many more markers need to be mapped first. A set of markers with the proper marker spacing can then be selected for efficient linkage studies. If new marker loci are first mapped physically, the number of marker combinations in linkage studies to determine the genetic distance can be reduced to combinations of markers that are located in the same chromosome region. During the process of saturating bovine genome with markers, monitoring the marker spacing will allow the identification of regions that are devoid of markers. Chromosome-specific libraries can then be used to saturate a specific region with marker loci while avoiding the use of new probes for a region that already carries sufficient markers.

It has now become generally accepted that the most efficient way to obtain the required number and distribution of polymorphic markers is to make use of DNA restriction fragment length polymorphisms (RFLPs) and of "variable number of tandem repeat" polymorphisms (VNTRs) in particular. DNA polymorphisms have been detected at high frequencies in humans. A linkage map of the entire human genome, based on the pattern of inheritance of about 400 marker loci, mostly RFLPs, has already been established (2). Maps consisting mostly of highly polymorphic human VNTR loci are also available for some human chromosomes (3–7). It is recommended that a major effort be made to produce marker maps in domestic animal species based on VNTRs (8). Strategies developed to isolate VNTRs (9,10) should also be applicable in isolating VNTRs in other species. Several groups have now started to search systematically for highly polymorphic loci in different domestic animal species.

METHODS TO ASSIGN LOCI TO PARTICULAR CHROMOSOMES

Physical mapping is carried out mainly by in situ hybridization or by the Southern blot or enzyme analysis of a somatic cell hybrid panel (11). However, since our experience in physical gene mapping is mostly restricted to in situ hybridization, and gene mapping using a somatic hybrid panel is described extensively elsewhere in this book, we will concentrate here on in situ hybridization.

IN SITU HYBRIDIZATION

In situ hybridization, as it is now routinely applied, involves hybridization of radiolabeled DNA probes to fixed metaphase chromosomes and subsequent visualization of the signal as silver grains after autoradiography. Hybridization in situ was originally developed by Gall and Pardue (12) and John et al. (13). This technique first allowed only localization of highly reiterated or amplified genes. Improvement in hybridization procedure, including the use of dextran sulfate in the reaction mixture, made possible the routine mapping of single copy genes. An example for in situ hybridization with a radiolabeled probe is shown in Figure 1, A and B. The probe used in this mapping experiment was a sequence of a bovine VNTR provided by M. Georges (Genemark Inc., Salt Lake City, Utah). The protocol followed for this experiment and generally applied in our laboratory for radioactive in situ hybridization is presented below. It is based on the method first described by Harper and Saunders (14).

PROCEDURES FOR IN SITU HYBRIDIZATION WITH RADIOLABELED PROBES

Preparation of Chromosomes

Chromosomes are prepared from leukocytes or from cultured fibroblast cells from male animals according to standard procedures. Usually several trials using blood cells from different animals are necessary to obtain preparations showing a satisfactory mitotic index. Consistently good results can be obtained when fibroblast cell lines are used that have been established from fetal tissue. Harvesting of these cells is by the mitotic shakeoff technique. The cell suspensions are dropped onto slides that are cleaned in 2 N HCl. Chromosome preparations are stored on slides in desiccated boxes at -20°C (usually several months).

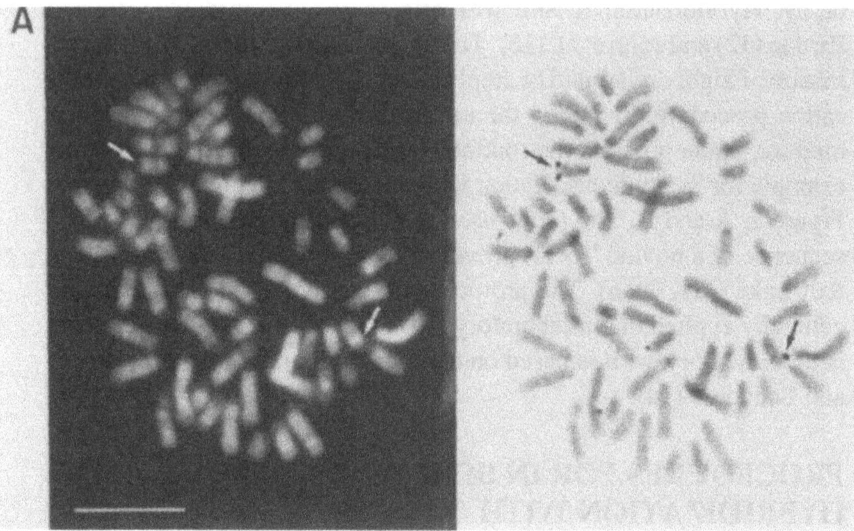

Figure 1 In situ hybridization with a radiolabeled probe of a bovine VNTR. (A) A bovine metaphase spread before (left, Q-banded) and after (right, Giemsa stained) hybridization and autoradiography for 8 days. The arrows point to chromosomes no. 21 and specific sites of labeling, respectively. (B) Histogram showing the distribution of autoradiographic silver grains in 107 metaphase spreads. Based on this silver grain distribution, the VNTR locus is assigned to bands 21q21–24.

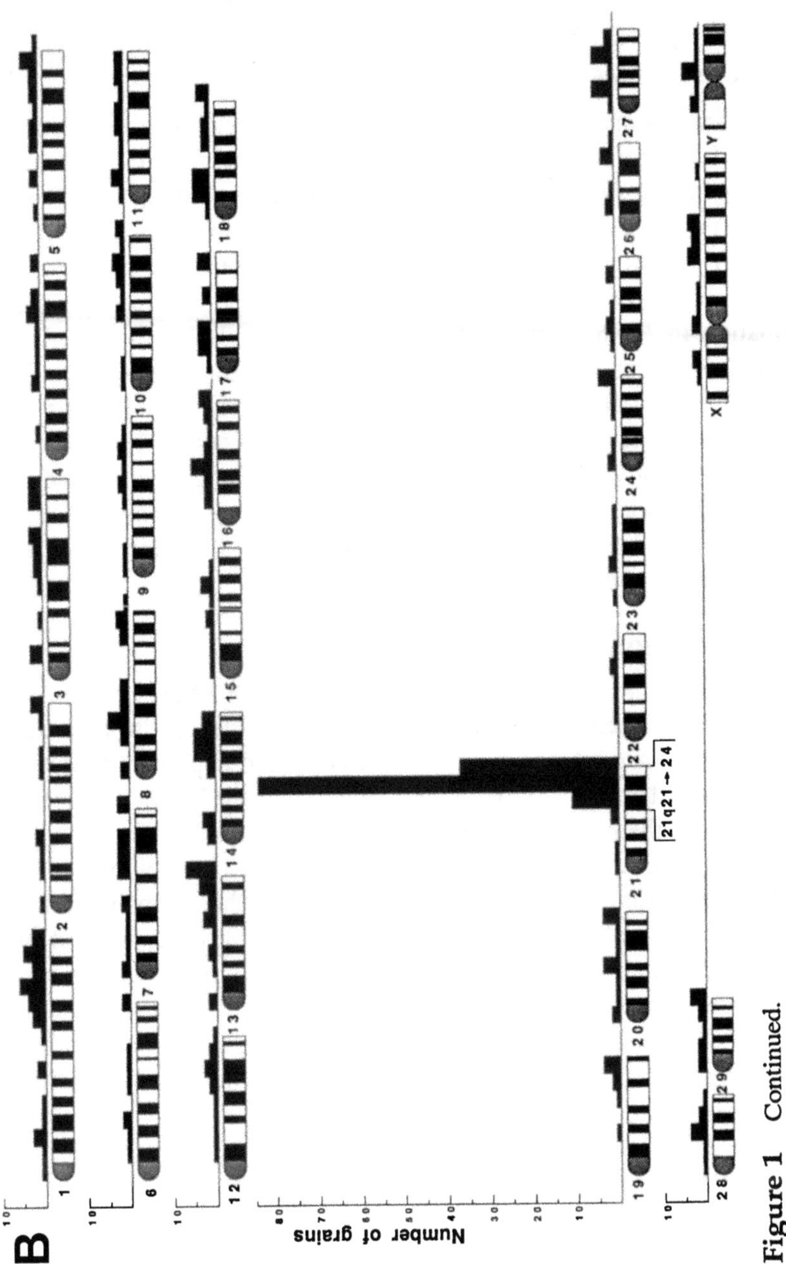

Figure 1 Continued.

Banding and Photography of Chromosomes

The chromosomes are Q-banded by dipping the slides in a solution of 0.005% quinacrine mustard (Fluka) in distilled water for 30 s. Thorough rinsing in deionized water after the staining is important to reduce background fluorescence. The slides are mounted in Sørensen's phosphate buffer (pH 6.8) and covered with cover slips without sealing. The metaphase spreads are viewed under a Zeiss microscope with the filter combination 48 77 06 or under a Leitz microscope with the filter combination 12/3. About 30 metaphase spreads are photographed per slide on a Kodak technical pan film 2415 and the position of the metaphase spreads is recorded. The films are developed in Kodak D19 developer. After the photographs have been taken, the coverslips are removed by rinsing the slides in deionized water and the slides are cleaned in several changes of xylene. The slides may then be stored in desiccated boxes at -20°C until required.

Denaturation of Chromosomes

The treatment of the chromosome preparation with RNase is optional; a clear improvement of the signal-to-noise ratio after in situ hybridization with RNase-treated preparations as opposed to preparations without treatment could not be observed. However, if the RNase treatment is carried out, the preparations are incubated for 1 h at 37°C in 100 ng/ml boiled RNase A (Sigma) in 2 x SSC (1 x SSC: 0.15 M sodium chloride, 0.015 M sodium citrate, pH 7.0). The slides are then dehydrated in an ethanol series of increasing concentrations (70%, 80%, 95%, 2 min each at room temperature). The dehydration step is also included when the RNase treatment is not performed. The chromosomes are denatured in 70% formamide (Fluka) in 2 x SSC for 2.5–5 min at 70°C. Immediately afterward, the slides are dehydrated in 70%, 80%, 90%, and 100% ethanol, 2 min each at room temperature.

Labeling of Probes

Labeling of 25 ng of insert DNA (often in low-geling temperature agarose) is performed with 75 pmol each of ^{3}H-dATP (ca. 180 Ci/mmol), ^{3}H-dCTP (ca. 70 Ci/mmol), and ^{3}H-TTP (ca. 120 Ci/mmol) (all tritiated nucleotides are from Amersham) and dGTP (500 pmol) in 20 μl containing the reaction mixture and Klenow Polymerase of the Random Primed DNA Labeling Kit of Boehringer. Specific activities of more than 1×10^8 dpm/μg probe DNA can be achieved using random primed labeling (15). In a few cases the whole plasmid is labeled. The labeled DNA is separated from unincorporated nucleoside triphosphates through a Sephadex G-50 column (Pharmacia) by elution with TNE [10 mM Tris-Hcl (pH8.0), 100 mM NaCl, 1 mM EDTA (pH 8.0)]. The labeled DNA is precipitated in the presence of 20 μg of alkaline-sheared salmon sperm DNA, recovered by centrifugation, and resuspended in the hybridization solution at a concentration of 50 ng/ml. The hybridization solution contains 2 × SSC, 10% (w/v) Dextran sulfate (Pharmacia), 50% formamide (Fluka), and 1% Tween 20 (Fluka).

Hybridization

The probe is first denatured in the hybridization solution for 5 min at 70°C and put on ice. Hybridization is carried out not later than 30 min after denaturation and dehydration of chromosomes by placing 50 μl of the hybridization mixture onto the slide. The slides are covered with coverslips of 24 × 60 mm. Not more than 1.25 ng of probe DNA is applied to one slide. More probe will cause an increase in background labeling. Hybridization is carried out in a moist chamber at 37°C for 16–20 h.

Washing

After hybridization, the coverslips are removed in 2 × SSC containing 50% formamide at 42°C. The slides are washed three times at 42°C in 2 × SSC containing 50% formamide for 5 min each and then three times at 42°C in 2 × SSC for 5 min each. Immediately after the wash step the slides are dehydrated in a series of 70%, 80%, and 90% ethanol, 2 min each at room temperature.

Autoradiography

For autoradiography, the slides are dipped in Ilford K2 emulsion (Ilford) diluted 1:1 in distilled water at 45°C and air-dried for 1 h. Handling of the emulsion is carried out under illumination through a Kodak

Safelight No. 2 filter. The slides are stored in light-tight, desiccated boxes at 4°. Each slide is kept in a separate box. The first slide of a 4-slide (per probe) standard experiment is developed after 1 week of exposure in Kodak D19 developer, diluted 1:1 in distilled water, rinsed briefly in deionized water and fixed for 5 min in Kodak fixer, rinsed in running deionized water for at least 20 min and air-dried. After autoradiography, the slides are stained with 5% Giemsa prepared in Sørensen's phosphate buffer (pH 6.8). If a large number of autoradiographic silver grains are associated with chromosomes, the remaining three slides are developed immediately afterward. If only a few grains can be seen, the next slide is developed after another week, and so forth. Optimum exposure time is usually 1-2 weeks. In the case of short probes (shorter than 1000 base pairs) the exposure can last up to 4 weeks.

Statistical Evaluation of Silver Grain Distribution

The distribution of autoradiographic silver grains is observed over the previously photographed chromosome spreads. The position of silver grains that are in contact with a chromosome is marked on prints of the metaphase spreads. The chromosome carrying label are identified based on the Q-banding pattern. The distribution of silver grains along the chromosomes is evaluated by plotting the position of silver grains on a histogram representing the haploid bovine genome. This plot is constructed by dividing a standard idiogram into 275 units scaled to the average diameter of a silver grain (0.55 μm). The standard idiogram of cattle has been finalized recently at the Second International Conference for the Standardization of Domestic Animal Karyotypes (15a). The standard idiogram of swine has been published (16). Standardization in other domestic species is still in progress. Statistical evaluation of the silver grain is based on the following empirical criterion: If the number of grains in a column is at least three times the number of grains in all other columns of the histogram (except in flanking columns), this column represents the signal. The number of grains in columns flanking this particular column is usually more than twice the number of grains in other columns. The gene locus is then assigned to the two bands below the peak column and the two flanking columns.

PROCEDURE FOR IN SITU HYBRIDIZATION WITH BIOTIN-LABELED PROBES

The routine procedure described above for radioactive in situ hybridization is very time consuming, because autoradiography may last several weeks and the evaluation of the silver grain distribution is a statistical procedure requiring analysis of up to 100 metaphase spreads. We have therefore recently started to apply biotin labeling to circumvent some of the problems. We were able to map a relatively short sequence in swine (Fig.2) after biotin labeling and detection of the signal with avidin-fluorescein isothiocyanat (FITC). The probe used in this experiment was a recombinant plasmid containing a porcine insert sequence (4.3 kb) of the major histocompatibility complex (SLA) encoding a class I antigen (17). Assignment of the signal to band p11 of porcine chromosome 7 corroborates earlier findings about the chromosomal location of SLA (18). The protocol for nonradioactive in situ hybridization given below is based on the procedure of Pinkel et al. (19), Lichter et al. (20), and Cherif et al. (21). Only steps that are different from the protocol for radioactive hybridization are described in detail below.

RNase and Proteinase Treatment

When nonradioactive in situ hybridization is performed, the RNase treatment, dehydration, and denaturation are always carried out as described above. After the second dehydration the slides are treated by proteinase K (Sigma) at a concentration of 60 ng/ml in 20 mM Tris-HCl (pH 7.5) and 2 mM $CaCl_2$ at 37°C for 7.5 min. The slides are then dehydrated in 70%, 80%, 90%, and 100% ethanol, 2 min each at room temperature.

Biotin Labeling

Labeling of DNA is either by the random priming method or by nick translation. Random priming is applied when only a small amount of insert DNA is available. Nick translation is usually applied when the whole plasmid is labeled. Random priming is performed using the Random Primed DNA Labeling Kit of Boehringer. Twenty-five nanograms of denatured insert DNA is labeled in a 20-μl reaction containing 1

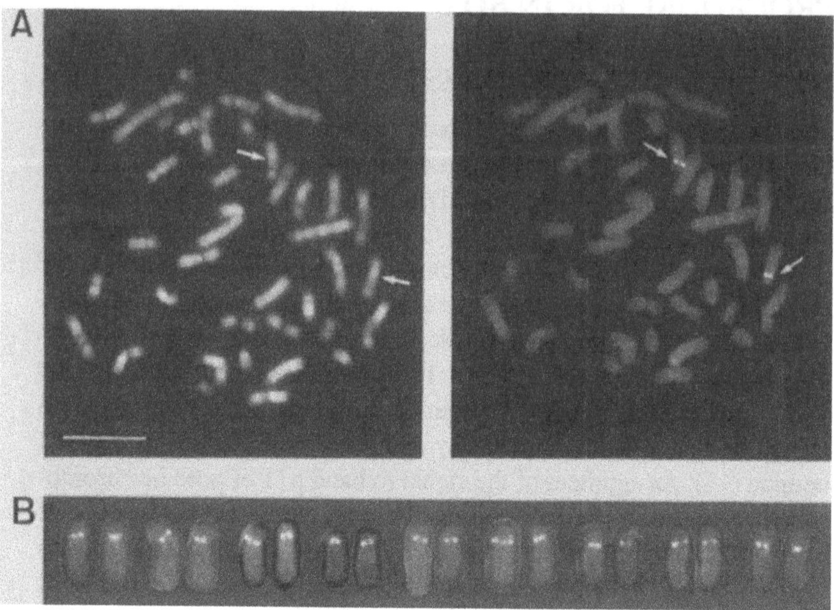

Figure 2 In situ hybridization using a biotinylated probe of the porcine major histocompatibility complex. (A) Porcine metaphase spread before (left, Q-banded) and after (right, propidium iodide stained) hybridization and three rounds of signal amplification. The arrows point to chromosomes no. 7 and the double spots of FITC-fluorescence, respectively, representing the hybridization signal. The bar represents 10 μm. (B) Pairs of homologous chromosomes no. 7 from 9 metaphase spreads, showing specific labeling on the short arm.

nmol Biotin-16-dUTP (Boehringer) and dCTP, dGTP, and dATP (1 nmol each, Boehringer) plus the reaction mixture and Klenow polymerase as specified by the manufacturer. In early experiments dATP was replaced by 70 pmol ^{3}H-dATP (Amersham, ca. 180 Ci/mmol) to identify the fractions that contained the labeled probe after purification on a Sephadex G-50 column (see above). In the more recent experiments, no tritiated nucleotide was added and the fractions containing the labeled probe were determined empirically. When nick translation is carried out, 200 ng of the insert containing plasmid is labeled using a nick-translation kit (Boehringer) in the following way: The DNA is biotinylated in a 20-μl reaction containing 1.2 nmol Biotin-16-dUTP (Boehringer), dATP, dGTP, dCTP (1.2 nmol each), nick translation buffer, and a mixture of DNA-polymerase I and DNase I as specified in the protocol accompanying the kit. Identification after Sephadex G-50 purification of fractions containing the labeled probe was initially by the addition of ^{3}H-dATP (75 pmol) instead of dATP and later empirically as described above. The further processing of the labeled DNA is also as described above. The biotinylated DNA is precipitated in presence of 20 μg alkali-sheared salmon sperm DNA and resuspended in the hybridization solution specified above. Hybridization is carried out as described for the radioactive in situ hybridization. However, 10–50 ng of probe DNA biotinylated by nick translation is applied to a slide.

Detection of the Signal

After hybridization, the slides are treated as described for the radioactive in situ hybridization. After the washings the slides are not dehydrated but immediately placed in equilibration buffer (4× SSC, 0.05% Tween 20). Slides may be stored in equilibration buffer for several days. The chromosome preparations are preincubated in 4 × SSC, 0.05% Tween 20, and 3% bovine serum albumin (Fraction V, Sigma) for at least 10 min at room temperature. For the incubation, 60 μl of the preincubation buffer containing 5 μg/ml of fluorescein avidin DSC (Vector) in 4 × SSC, 0.05% Tween 20, and 3% bovine serum albumin is placed on a slide, covered by a coverslip, and incubated in a moist chamber for 45 min. The coverslip is then carefully removed in 4 × SSC, 0.05% Tween 20, and the slides are washed three times in 4 × SSC, 0.05% Tween 20 (3 min each). The chromosome preparations are then briefly rinsed in deionized water and stained for 90 s in propidium iodide at a concentration of 2 μg/ml in distilled water and rinsed again in deionized water. Sixty microliters of antifade solution (22) is applied to each slide, and then the slides are covered with a coverslip. Metaphase spreads of the pictures taken earlier are now relocated under a Leitz fluorescence microscope. Fluorescein and propidium iodide can be excited simultaneously at 450–490 nm (Leitz combination 12/3). If a signal is present, it appears as strong yellow-green spots on the red chromosomes. In most cases, however, several rounds of amplification are necessary to obtain a sufficiently strong signal.

Signal Amplification

This is accomplished after careful removal of the coverslip and rinsing in 4 × SSC, 0.05% Tween 20, by placing 60 μl of 5 μg/ml biotinylated goat anti-avidin D (Vector) in 4 x SSC, 0.05% Tween 20, and 3% bovine serum albumin on each slide and covering them with coverslips. The incubation is for 45 min at room temperature in a moist chamber. The coverslips are then removed and the slides are washed as described above for the washings after incubation with fluorescein avidin. Another incubation with fluorescein with subsequent washing and staining is also carried out as described above. We have accomplished up to four amplification rounds without encountering significant background problems. A strong signal appears as brilliant spots on both chromatides. Often both homologous chromosomes exhibit doublets of brilliant spots. Background, if any, appears as much less brilliant single spots. Using a film (Agfa Ortho 25 Professional) that is extremely green sensitive and rather insensitive to red allows the documentation of metaphase spreads on black and white prints.

INCREASED EFFICIENCY AND RESOLUTION OF NONRADIOACTIVE IN SITU HYBRIDIZATION VERSUS RADIOACTIVE HYBRIDIZATION

The significant reduction of background labeling that can be achieved when applying nonradioactive in situ hybridization allows the number of metaphase spreads to be analyzed to be reduced to 5–10. Thus, nonradioactive in situ hybridization will make physical mapping more efficient. Another advantage of biotin labeling is the increased mapping resolution that can be achieved through the direct visualization of the signal by avidin-FITC. The increased resolution of nonradioactive hybridization is demonstrated in Figure 3, A and B. An anonymous sequence specific for the Y chromosome has first been regionally mapped on the Y chromosome by radioactive in situ hybridization (23). After

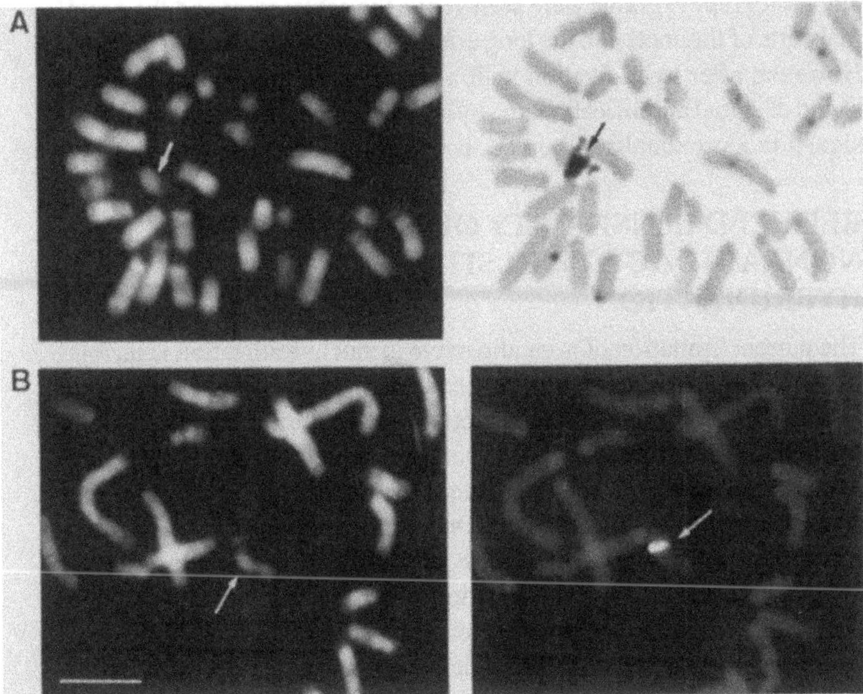

Figure 3 Radioactive and nonradioactive in situ hybridization with a Y chromosome-specific sequence. The arrows point to the centromeres of the Y chromosomes and to the specific signals, respectively. The bar represents 5 μm (A) A partial metaphase spread before (Q-banded, left) and after (Giemsa stained, right) hybridization and autoradiography for 24 h. (B) A partial metaphase spread before (Q-banded, left) and after (propidium iodide stained, right) hybridization and two rounds of signal amplification. The signal is originating from band Yp12.

autoradiography, grains were seen above the centromere and the proximal parts of the short and the long arms of the Y chromosome (Fig. 3A). However, after hybridization with a biotinylated probe, and after one round of signal amplification, the hybridization site is unequivocally assigned to a proximal band on the p arm (Fig. 3B).

REDUCED SENSITIVITY OF NONRADIOACTIVE IN SITU HYBRIDIZATION

The current limitation of nonradioactive in situ hybridization is its relatively low sensitivity. That is, probes that are smaller than 1 kb cannot be routinely assigned by nonradioactive in situ hybridization, especially when the avidin–FITC signal detection technique is used. The enzymatic detection of biotinylated sequences seems to result in an increased sensitivity of the method. One of these methods involves incubation with strepavidin–peroxidase and addition of diaminobenzidin (DAB) as substrate (24). Precipitation of DAB at the site of hybridization can be visualized with reflection contrast microscopy (25). Localization of the hybridization signal, however, is less precise with the enzymatic detection method than with the avidin–FITC method.

COSMID CLONES AS PROBES FOR NONRADIOACTIVE IN SITU HYBRIDIZATION

The strategy that has yielded a large number of highly polymorphic VNTRs in humans consisted of screening a cosmid library with suitable probes (10). The same strategy will most likely also lead to detection of numerous VNTRs in domestic animal species. Cosmid clones will then be available as probes for in situ hybridization. However, each cosmid probe will probably contain repetitive sequences, which can obscure the specific signal resulting from the unique parts of probes. These repetitive sequences can be eliminated through the addition of competitor DNA allowing the use of cosmid probes for in situ hybridization just like single–copy probes (26). Because an average cosmid contains more than 10 kb of unique sequences, a detectable signal will be present even after hybridization with biotinylated probes. Thus, cosmid clones that are candidates for detecting VNTRs may first be chromosomally assigned, and only in the case of an interesting localization they may further be characterized.

COMBINATION OF PANEL MAPPING AND NONRADIOACTIVE IN SITU HYBRIDIZATION

Assignment of gene loci to the physical map is possible after analysis of a panel of hybrid lines. This approach, however, does not allow the regional assignment of a gene on a chromosome, unless hybrid lines are available with specific chromosomal rearrangements. Another drawback of panel mapping in cattle are the difficulties involved in identifying the cattle chromosomes in hybrid cells. Mapping of suitable members of groups of syntenic genes by in situ hybridization seems the only way to assign syntenic groups to specific chromosomes. However, it can be expected that all syntenic groups will be chromosomally assigned in the near future using this indirect approach. Combining the methods of in situ hybridization and panel mapping may then be the most efficient way to map marker loci: Southern blot analysis of a panel of hybrid lines will be a fast way to assign a new marker to a synteny group and respective chromosome, while the regional localization on the chromosome is achieved afterward by nonradioactive in situ hybridization. Less than five metaphases could be sufficient to perform a regional assignment once the chromosome is know.

CONCLUSION

The establishment of marker maps in domestic species is of significant interest as a basis for systematic dissection of the genome with regard to economically important traits. Saturating of genomes with marker loci requires permanent monitoring of the achieved coverage. We propose physical mapping of all new marker loci to study their distribution on the chromosomes. Linkage studies to evaluate the genetic distances between marker loci will be simplified since the order of loci is established beforehand on the physical map. Among the methods for the chromosomal assignment of marker loci, nonradioactive in situ hybridization may be the most efficient approach, perhaps in combination with the analysis of somantic hybrid panels. In any case, physical gene mapping will become an integral part of what could be the "marker initiative" in domestic animal species.

ACKNOWLEDGMENTS

We thank G. Stranzinger for his constructive arguments and K. Wüthrich for technical assistance. Probes were generously provided by M. Georges and D. Singer. Our work was made possible by grants from the Federal Institute of Technology and the Swiss National Science Foundation.

REFERENCES

1. Ruddle FH, Fries R. Mapping genes in domesticated animals. In: Genetic engineering of animals: an agricultural perspective. Basic Life Sci 1986; 37:39-57.
2. Donis-Keller H, Green P, Helms C, Cartinhour S, Weiffenbach B, Stephens K, Keith TP, Bowden DW, Smith DR, Lander ES, Botstein D, Akots G, Rediker KS, Gravius T, Brown VA, Rising MB, Parker C, Powers JA, Watt DE, Kauffman ER, Bricker A, Phipps P, Muller-Kahle H, Fulton TR, Ng S, Schumm JW, Braman JC, Crooks SM, Lincoln SE, Daly MJ, Abrahamson J. A genetic linkage map of the human genome. Cell 1987; 51:319-337.
3. Lathrop M, Nakamura Y, O'Connell P, Leppert M, Woodward S, Lalouel J- M, White R. A mapped set of genetic markers for human chromosome 9. Genomics 1988; 3:361-366.
4. Nakamura Y, Lathrop M, Bragg T, Leppert M, O'Connell P, Jones C, Lalouel J-M, White R. An extended genetic linkage map of markers for human chromosome 10. Genomics 1988; 3:389-392.
5. Nakamura Y, Lathrop M, O'Connell P, Leppert M, Barker D, Wright E, Skolnick M, Kondoleon S, Litt M, Lalouel J-M, WHite R. A mapped set of DNA markers for human chromosome 17. Genomics 1988; 2:302-309.
6. O'Connell PO, Lathrop GM, Leppert M, Nakamura Y, Müller U, Lalouel J-M, White R. Twelve loci form a continuous linkage map for human chromosome 18. Genomics 1988; 3:367-372.
7. Nakamura Y, Lathrop M, O'Connell P, Leppert M, Lalouel J-M, White R. A primary map of 101 DNA markers and two serological markers for human chromosome 19. Genomics 1988; 3:67-71.

8. Fries R, Beckmann JS, Georges M, Soller M, Womack J. The bovine gene map. Anim Genet 1989; 20:113–139.
9. Wong A, Wilson V, Jeffreys AJ, Thein SL. Cloning a selected fragment from a human DNA fingerprint: isolation of an extremely polymorphic minisatellite. Nucl Acids Res 1986; 14:4505–4616.
10. Nakamura Y, Leppert M, O'Connell P, Wolff R, Holm T, Culver M, Martin C, Fujimoto E, Hoff M, Kumlin E, Whyte R. Variable number of tandem repeat (VNTR) markers for human gene mapping. Science 1987; 235:1616–1622.
11. Ruddle FH. A new era in mammalian gene mapping: somantic cell genetics and recombinant DNA methodologies. Nature 1981; 294:115–120.
12. Gall JG, Pardue ML. Formation and detection of RNA-DNA hybrid molecules in cytological preparations. Proc Natl Acad Sci USA 1969; 63:378- 383.
13. John HA, Birnstiel ML, Jones KW. RNA-DNA hybrids at the cytological level. Nature 1969; 223:234–238.
14. Harper ME, Saunders GF. Localization of single copy DNA sequences on G-banded human chromosomes by in situ hybridization. Chromosoma 1981; 83:431- 439.
15. Feinberg AP, Vogelstein B. A technique for radiolabeling DNA restriction endonuclease fragments to high specific activity (addendum). Analyt Biochem 1984; 137:266–267.
15a. ISCNDA (1985). International System for Cytogenetic Nomenclature of Domestic Animals, Di Berardion D, Hayes H, Fries R, Long S (eds.). Cytogenet Cell Genet 1990; 53:65–79.
16. Gustavsson I. Standard karyotype of the domestic pig. Committee for the Standardized Karyotype of the Domestic Pig. Hereditas 1988; 109:151–157.
17. Singer DS, Lifshitz R, Abelson L, Nyirjesy P, Rudikoff S. Specific association of repetitive DNA sequences with major histocompatibility genes. Mol Cell Biol 1983; 3:903-913.
18. Rabin M, Fries R, Singer D, Ruddle FH. Assignment of the porcine major histocompatibility complex to chromosome 7 by in situ hybridization. Cytogenet Cell Genet 1985; 39:206–209.
19. Pinkel D, Straume T, Gray JW. Cytogenetic analysis using quantitative, high-sensitivity fluorescence hybridization. Proc Natl Acad Sci USA 1986; 83:2934–2938.
20. Lichter P, Cremer T, Borden J, Manuelidis L, Ward DC. Delineation of individual human chromosomes in metaphase and interphase cells by in

situ suppression hybridization using recombinant DNA libraries. Hum Genet 1988; 80:224-234.

21. Cherif D, Bernard O, Berger R. Detection of single-copy genes by nonisotopic in situ hybridization on human chromosomes. Hum Genet 1989; 81:358-362.
22. Johnson GD, Davidson RS, McNamee KC, Russell G, Goodwin RD, Holborow EJ. Fading of immunofluorescence during microscopy: a study of the phenomenon and its remedy. J Immunol Meth 1982; 55:231-242.
23. Perret J, Shia Y-C, Fries R, Vassart G, Georges M. A polymorphic satellite sequence maps to the pericentric region of the bovine Y chromosome Genomics 1990; 6:482-490.
24. Ambros PF, Karlic HI. Chromosomal insertion of human papillamavirus 18 sequences in HeLa cells detected by nonisotopic in situ hybridization and reflection contrast microscopy. Hum Genet 1987; 77:251-254.
25. Landegent JE, Jansen in de Wal N, Ploem JS, van der Ploeg M. Sensitive detection of hybridocytochemical results by means of reflection-contrast microscopy. J Histochem Cytochem 1985; 33:1241-1246.
26. Landegent JE, Jansen in de Wal N, Dirks RW, Baas F, van der Ploeg M. Use of whole cosmid cloned genomic sequences for chromosomal localization by nonradioactive in situ hybridization. Hum Genet 1987; 77:366-370.

9

Data Analysis for Linkage Studies

Mark Lathrop

Centre d'Etude du Polymorphism Humain, Paris, France

P. Cartwright

University of Utah, Salt Lake City, Utah

S. Wright

Native Plants Inc., Salt Lake City, Utah

Y. Nakamura

Cancer Institute, Tokyo, Japan

Michel Georges

Genmark Inc., Salt Lake City, Utah

INTRODUCTION

Recent advances in molecular biology have made possible extensive linkage investigations in many species. In humans, the genes responsible for several genetic diseases have been isolated from knowledge of their genetic locations obtained from the linkage studies; these include the genes responsible for Duchenne muscular dystrophy (Monaco et al., 1986) and cystic fibrosis (Riordan et al., 1989). Although systematic efforts to develop polymorphic markers and construct genetic linkage maps of chromosomes in humans have advanced furthest, similar endeavors are now underway for many species of domestic plants and animals. The purpose of this paper is to provide an introduction to the data management and analysis needed to conduct these linkage studies.

DETECTION OF LINKAGE

The goal of linkage studies is to assign the genetic factors responsible for a trait phenotype to specific chromosome regions by examining their cosegregation with marker loci. Most modern linkage studies have relied on the method of maximum likelihood as the principle technique for statistical estimation of recombination rates and evaluation of linkage (see Ott, 1985). When two codominant loci are characterized in a backcross or certain other simple phase-known matings, the likelihood function is a binomial of the form

$$\Theta^r(1 - \Theta)^{n-r}$$

where r is the number of recombinants, and n is the number of informative meioses. In this case, the maximum likelihood estimate is $\Theta = r/n$, i.e., the observed recombination frequency, if $r/n \leq 0.5$. For, $r/n > 0.5$, it is usual to estimate the recombination frequency as $\Theta = 0.5$. A test of no linkage (i.e., $\Theta = 0.5$) against the alternative $\Theta \leq 0.5$ can be obtained from the likelihood ratio statistic:

$$LR = \frac{\left[\Theta^r(1 - \Theta)^{n-r}\right]}{0.5^n}$$

By convention, the likelihood ratio test is usually presented in the form of the lod score, or lod_{10} (LR), although 2 ln (LR) would be more convenient from a statistical viewpopint as the latter has a chi-square distribution with 1 degree of freedom under the null hypothesis of no linkage.

In human genetics, a lod score >3 (equivalent to $0.001 > p > 0.0001$ in a one-sided test) is accepted as significant evidence of linkage (Morton, 1955; Smith, 1959). The prior probability of linkage between two autosomal loci has been used to justify this stringent critical value; since the prior probability that randomly chosen loci will be on the same autosomal chromosome is approximately 1/20 in humans, the frequency of nonsyntenic loci pairs among loci comparison leading to a lod score >3 is expected to be between 1% and 0.1%. Although the prior probability of linkage can be obtained with greater accuracy with more sophisticated distribution models, a critical value of 3 is reasonable as an initial indicator for linkage can be obtained rapidly through characterization of other markers or additional meioses.

Maximum-likelihood estimates of the recombination estimate can be found from direct counts of recombinant meioses in a few circumstances. However, the majority of linkage investigations involve complex pedigrees, phenotypic traits with reduced penetrance or quantitative values, unknown phase, and missing data. Generally, numerical algorithms and computer programs are required for evaluation and maximization of the likelihood. Widely available programs for linkage analysis of general pedigrees and phenotype include LIPED (Ott, 1974), LINKAGE (Lathrop and Lalouel, 1984), and MENDEL (Lange et al., 1988). The following sections examine some of the issues involved in linkage analysis with such programs.

TRAIT PHENOTYPES

Trait phenotypes are usually indirect measures of the genotype of interest that are often presented as categorical data or quantitative measurements. Many computer programs for linkage analysis consider categorical variables as binary responses, e.g., affected or nonaffected for a genetic disease phenotype. Often the binary responses are denoted as 1 (normal or nonexpressed phenotype) or 2 (disease or expressed phenotype), and 0 is reserved to indicate that the phenotype is unknown. It is possible to allow recessive, dominant, or other modes of inheritance by appropriate specification of the penetrance or the probability of expression of the phenotype for each of the trait genotypes. Typically, two allelic variants, D and d, will be assumed (e.g., normal and disease alleles, respectively), so that the three genotypes DD, Dd, and dd are possible at the trait locus, and the corresponding penetrances are denoted as f_{DD}, f_{Dd}, and f_{dd}. The trait is said to be completely penetrant and recessive if $f_{DD} = f_{Dd} = 0$ and $f_{dd} = 1$; it is completely penetrant and dominant if $f_{DD} = 0$ and $f_{Dd} = f_{dd} = 1$. Small, nonzero values can be introduced to allow for phenocopies; the genotype is incompletely penetrant if the corresponding penetrance is different from 0 or 1. Incomplete expression for dominant or recessive traits is common and can be accounted for by adjusting the corresponding penetrances to be <1.

Penetrances can be modified by different factors such as age, sex, and environmental variables. These can be accounted by defining a discrete liability class associated with each individual. An example is an age class, where ages are divided into broad classes with approximately equal penetrance throughout the class. Categorical phenotypes with >2 responses can also be analyzed by direct extension to account for a greater number of penetrance classes, or as a quantitative variable where the phenotype is expressed as observed/maximum response.

Quantitative phenotypic variables are sometimes dichotomized to obtained a single binary response, but this often results in a loss of information for linkage studies. On the other hand, analysis of a quantitative variable requires assumptions on the distribution of the phenotype; a quantitative phenotype is usually modeled by assuming that the trait distribution is a mixture of normal variables with different genotype-specific means, and common residual variance (variance about the means); square-root, logarithmic, or other transformations may be required to obtain the appropriate distribution. Several variables can be considered as a mixture of multivariate normal distributions; similarly, it is possible to measure phenotypes simultaneously by qualitative and quantitative variables (see below).

If the assumption of equal residual variances is relaxed, the parameters of the mixture distribution are no longer identifiable in the statistical analysis [see Everitt and Hand (1981, chapter 2) for a discussion in a general context]. Data on known carriers or noncarriers of a gene may sometimes provide estimates of different variances for each genotype, when the assumption of a common variance is held to be inappropriate.

In many instances, several genetic factors may influence the phenotypic expression. One approach to analysis of quantitative multifactorial genetic data is to allow for a single major gene and residual familial correlation that follows an additive genetic model, under the so-called "mixed" genetic model [see Hasstedt (1982) for a development of the mixed model with linked markers]. An alternative approach with "regressive models" has recently been described (Bonney et al., 1988). In some experimental designs, such a backcross or F2 panels formed by starting from completely inbred parental lines, the structure of the parental population allows simplification for multifactorial traits [see

Lander and Botstein (1989) for a recent discussion of one possible approach in this situation].

MARKER PHENOTYPES

Most marker systems are codominant, with corresponding marker phenotypes that can be represented as a unique genotype composed of two alleles. Sometimes it is necessary to consider more complicated systems that lack a one-to-one correspondence of marker genotypes and their phenotypes. Such data can be represented as a factor-union system in which phenotypes and alleles are coded as the presence or absence of a set of factors. A genotype (i.e., allelic pair) is said to be compatible with a phenotype if the union of the allelic factor sets is equal to the phenotype set. For example, a recessive/dominant marker phenotype can be defined as the presence or absence of a single factor, such as a band recognized by hybridization of a minisatellite sequence, and the two alleles are defined by a set in which the factor is absent and a set in which it is present.

EXAMPLE: MUSCULAR HYPERTROPHY in CATTLE

The following example illustrates some of the points raised in the above discussion. In a recent study, Georges et al. (1990) have studied linkage between minisatellite bands detected as fingerprint patterns and the gene responsible for muscular hypertrophy ("double-muscled" phenotype) in cattle. Date were collected on several half-sib pedigrees, each consisting of a single founder bull mated to many dams; in most instances, the pedigree contained a single offspring for each dam.

A single recessive gene (denoted "mh") is responsible for muscular hypertrophy (Hanset and Michaux, 1985). Animals were classified as having a double-muscled or conventional phenotype after visual examination. Plasma and red blood cell concentrations of creatine and creatinine were measured to improve the phenotypic classification. Two sires were classified as conventional animals based on biochemical data and extensive examination of muscular development by several trained observers. The frequency of double-muscled animals among the offspring of these two bulls confirmed their status as heterozygotes.

Linkage analysis between minisatellite markers and the trait locus was restricted to these two pedigrees. Fingerprint patterns were determined in the parents and offspring with five different probes, and bands were scored as present or absent. Since heterozygotes and homozygotes for the presence of a band could not be distinguished, each system was considered as a recessive/dominant marker phenotype. Note that only those bands present in the bull are informative for linkage to the trait.

Analysis with the LINKAGE programs (Lathrop and Lalouel, 1984) proceeded in the following way. Minisatellite phenotypes were entered into a computerized database using a factor–union system with two factors; the first factor indicated the presence or absence of the band; the second factor indicated that the band had been interpreted. Thus, the possible phenotypes were 1–1 (band present), 0–1 (band absent), and 0–0 (not interpreted). The second factor is introduced here to indicate the difference between an absent band and noninterpreted experimental result. The alleles were coded as 1–1 and 0–1; thus the phenotype 1–1 could be contained as the union of either alleles 1–1 and 1–1 or alleles 1–1 and 0–1. Gene frequency estimates for each band were obtained by maximum-likelihood methods prior to the linkage analysis.

Animals were scored on the basis of the visual classification as 2 ("double-muscled", i.e., genotype mh/mh), 1 (conventional, i.e., genotype mh/+ or +/+, or 0 (not interpreted). Possible errors in the visual classification of dams and offspring were accounted for by allowing variable penetrances for the different genotypes. For example, we allowed for errors in the classification of double-muscled animals by estimating the penetrance $f_{mh/mh}$ (< = 1). Similarly, we allowed for errors in the classification of conventional animals by estimating f_{mh} and $f_{+/+}$ (> = 0). Since the bulls had been assigned a conventional phenotype after extensive investigation, they were assigned to a separate liability class in which the classification was assumed to be completely accurate (i.e., full penetrance). Joint maximum-likelihood estimation of the penetrance and gene frequency of the mh allele gave $f_{mh/mh}$ =0.8, $f_{mh/+}$ – 0, $f_{+/+}$ = 0, and p_{mh} (gene frequency) = 0.34.

Quantitative measurements of plasma and red blood cell levels of creatine and creatinine were analyzed as mixtures of normal distributions; visual classification of the animals was taken into account simultaneously. Plasma creatine exhibited the largest discrimination between genotypes, with estimated genotypic means of 3333 for +/+, 2597 for +/mh, and 1073 for mh/mh and an estimated common standard deviation of 659. Estimates of the gene frequency and penetrances were virtually unchanged by inclusion of the quantitative data.

Parameter estimates obtained as described above were used for the linkage analysis. Estimation of the recombination between the minisatellite bands and the visual phenotype led to discovery of a band (denoted EBA3) in one pedigree (57 dam offspring) pairs with a maximum lod score of 2.4 at $\Theta = 0$. All other lod scores in the two pedigrees were < 1.6. When the quantitative variables were included in the analysis, either singly or in combination, the lod scores ranged from 2.8 to 4.1. Lod scores < 3 were found only when considering plasma creatine. Examination of the data revealed a single offspring that had not received the EBA3 band for which the value for this variable was consistent with the mh/mh genotype. Other values for this individual were normal, and it was not given a visual classification because of an ambiguous external morphology.

PEDIGREE DATA

A pedigree structure for genetic studies is usually defined by a series of family and individual identifiers, most of which serves as pointers to move through the pedigree for likelihood calculations. The minimal set of information that is necessary for this purpose are: pedigree; individual, father, and mother identification numbers; and sex. It is usually necessary to include some individuals fro whom marker and trait phenotype data are not available to define pedigree links.

An important issue in the planning of linkage studies is evaluation of the potential to detect linkage with available pedigrees. Although the power to detect linkage can be determined by exact numerical calculations for some study designs, such as backcross, F_2, or recombination inbred panels of fixed size, simulation is often the only feasible solution. Recently, methods have been described for simulation studies of general pedigrees and phenotypic traits (Ploughman and Boehnke, 1989; Ott, 1989). Examples of the application of one of these methods to determine the power of specific pedigree structure are given by Lathrop and Lalouel (1990). An implementation of the method due to

Ott (1989) within the LINKAGE programs is available upon request from D. Weeks (Dept of Genetics, Columbia University, New York).

MULTILOCUS METHODS

Many linkage studies require analysis of several loci simultaneously. Multilocus linkage analysis requires methods for simultaneous estimation of gene order and genetic distance between loci (Lathrop et al., 1984, 1985). Efficient methods for evaluation of the likelihoods and parameter estimation for different types of data have been developed (Lathrop et al., 1986; Lathrop and Lalouel, Clayton, 1986; Lander and Green, 1987; Lander et al., 1987). In certain experimental designs, such as crosses from inbred lines leading to F_2 and backcross progeny, the likelihood equations for multilocus linkage analysis are particularly simple (Lathrop and Lalouel, 1990). Most approaches to multilocus analysis assume lack of interferences in crossing-over;; statistical investigations have shown that this assumption should not have significant consequences on the inference of gene order and recombination rates even in the presence of substantial interference.

The evidence for two different gene orders is usually compared by calculation of the odds ratio, i.e., the ratio of the maximum of the likelihoods under the two orders. Although no simple statistical test exists for this comparison of one order to another, odds of 100:1 or 1000:1 against are usually considered reliable for rejection of the order with smaller likelihood.

DATA VERIFICATION

Data verification is essential in any genetic linkage study, as genotype errors can lead to apparent recombinants that bias estimates of recombination rates and affect the determination of gene order. Consistency of the pedigree relationships should be verified with a highly polymorphic marker, such as a minisatellite sequence. The data on all obligate and probable recombinants should be verified for closely linked markers (< 10% recombinants), particularly when multiple recombinants occur within the same family closely linked markers in the same meioses should be identified for data verification once the maximum-likelihood gene order has been determined, as double or triple recombination

```
         1  3         5  6           2  4          6  6
         2  2         1  1           2  2          1  2
         1  4         3  3           4  5          2  4
         1  2         2  2           1  2          2  2


              3  6                         4  6
              2  1                         2  1
              4  3                         5  4
              2  2                         1  2


|3||4|  |3|\6/  \6/|4|  \6/\6/  |3|\6/  |3||4|  |3||4|  \6/|4|
|2||2|  |2|\1/  \1/|2|  \1/\1/  |2|\1/  |2||2|  |2||2|  \1/\1/
|4||5|  |4|\4/  \3/|5|  \3/\4/  |4|\4/  \3/|5|  |4||5|  \3/|5|
 2 |1|   2 \2/   2 |1|   2 \2/   2 \2/   2 |1|   2 |1|   2 |1|

                                           ^               ^
```

Figure 1 Data from four closely linked markers on chromosome 14, detected by probes MHZ9, MLJ14, CMM62, AW101 (given in the most probable genetic order) typed in one of the CEPH reference families. Genotypes are given for the four grandparents, two parents, and eight children. Alleles are number sequentially at each locus. In this example, the parental phase can be determined uniquely from the grandparent genotypes. Children genotypes are indicated in the most probable phase, with the chromosome inherited from the father listed first. The most likely origin of an allele is indicated as | | for the first parental chromosome, and \ / for the second parental chromosome, when the locus is informative. Recombinant chromosomes (indicated ^) are identified by alleles with different parental chromosome origins.

events are likely to be due to data errors. Computer programs for identification of recombination events are essentially for efficient data verification; programs for this purpose will be made as part of the LINKAGE package in the near future.

Figure 1 illustrates the power of multilocus methods for detection of genotype error. The data are from four closely linked markers on chromosome 14, detected by probes MHZ9, MLJ14, CMM62, AW101 (given in the most probable genetic order) typed in one of the CEPH reference families (see next section). The sex-averaged recombination distances between adjacent markers are 0.02, 0.02, and 0.07 (Nakamura et al., 1989). An error has been introduced in the genotype at the second locus in the last child, which has changed from 1,2 to 1,1 to mimic a typical misinterpretation of allele. The most probable phase distribution of alleles was then calculated using the LINKAGE programs; the results shown in Figure 1 led to identification of two recombinant individuals. For the last child, the potential error at the second locus can be seen immediately since the data can only be explained if a double recombination event is invoked in each of the flanking intervals. Reinterpretation should be envisioned in the situation rather than elimination of the data, since double recombination events are theoretically possible even in short intervals.

GENETIC MAPS FROM REFERENCE PANELS

The efficiency of linkage studies of human genetic disease and other phenotypic traits in different species is greatly enhanced by the availability of detailed genetic maps. In humans, primary genetic linkage maps (10–20% recombination between adjacent markers) are now available for most chromosomes, and high-resolution maps 1–5% recombination) are being developed (reviewed by Lathrop and Nakamura, 1990); similar maps are being developed for many other species. Primary maps permit rapid localization of the gene responsible for any mendelian trait after characterization of selected markers, if a sufficient number of informative meioses can be obtained for linkage studies.

An effective strategy for construction of genetic linkage maps is to form a reference panel of families or meioses from experimental crosses to allow investigation of linkage between a large number of marker loci. In humans, immortalized cell lines have been established to provide a permanent source of DNA from large, two or three-generational nuclear families (White et al., 1985). Grandparents are included, when possible, to verify parental genotypes and to provide phase information. DNA from 40 families selected from this panel and other sources is distributed by the Centre d'Etude de Polymorphism Humain (CEPH) in Paris to laboratories working in the field (Dausset et al., 1990). A common reference panel allows rapid construction of genetic maps since genotypic data from different laboratories can be combined.

Widely distributed panels of recombinant inbreds (RI panels) serve a similar purpose for linkage studies in other organisms, particularly in maize (Burr et al., 1988) and the mouse (Taylor, 1989). Although RI lines provide a potentially unlimited source of DNA, they have the disadvantage of being difficult to construct and maintain in large numbers. Since many possibilities for recombination occur when the lines are formed, significant linkage is usually seen only for closely linked markers. Backcross or F_2 panels are often used as an alternative to RI lines, but they have the disadvantage of limited quantities of DNA.

CONSTRUCTION OF GENETIC LINKAGE MAPS

Construction of linkage maps from genotypes on a reference panel usually proceeds by the following steps:

1. Genotypes are obtained for a number of loci with known chromosomal localizations to serve as reference points.
2. Other markers are characterized on the panel and localized to particular chromosome regions by pairwise linkage studies with previously assigned loci.
3. An initial, or trial, order is found for the loci assigned to a chromosome, either from the pairwise recombination estimates and lod scored or by minimization of obligate crossovers.
4. The evidence for or against different alternative orders is evaluated by multilocus likelihood methods to obtain an order with maximum support (the most likely locus order examined).
5. Distances between the loci are estimated under the order with maximum support.

Many methods are available as trial map procedures for step 3 [see the reviews in White and Lalouel (1987) or Lathrop and Lalouel (1990)]. References to methods for efficient multilocus likelihood evaluation for reference panels were given above. Since the total number of possible gene orders is extremely large, even when moderate numbers of marker loci have been localized to a particular chromosome, usually it is not possible to examine all orders in stage 4 (Lalouel et al,. 1986). Algorithms for selecting subsets of orders to examine are described in

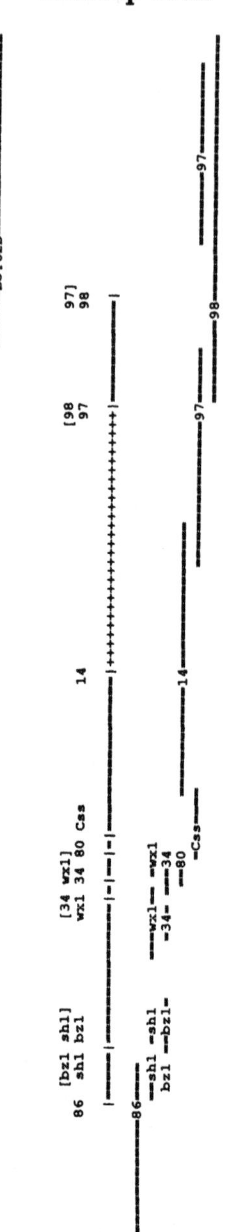

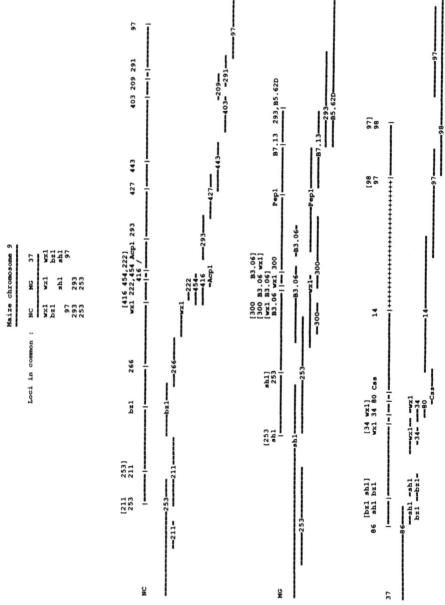

Figure 2

Lathrop et al. (1988a) and Donis-Keller et al., 1987). As described above, the data should be reverified after the final gene order has been obtained to eliminate possible errors leading to multiple recombination events. Therefore, it is usually necessary to repeat the analysis of data twice to produce a single chromosome map.

In addition to the order of markers and distance between loci, the representation of a linkage map should include information on statistical support or confidence in the inference. Standard methods for representing statistical support include: (1) reporting all alternative orders with odds within 100:1 or 1000:1 of the best supported order; (2) reporting odds against the inversion of adjacent loci; and (3) approximate confidence intervals for the placement of each locus [see Lathrop and Nakamura (1990) for examples and further references].

EXAMPLE: LINKAGE MAPS IN MAIZE

The following example is intended to illustrate the representation of genetic linkage maps. A total of 31 markers on maize chromosome 9 were characterized in two F_2 panels, denoted NC and 37 (containing 187 and 47 progeny, respectively), and BC_1 panel, denoted MG (containing 70 progeny) [see Helentjaris et al. (1986) for details]. Because of the paucity of common markers that had been characterized in all panels (see Fig. 2), we constructed separate genetic linkage maps for each panel. Recombination estimates and lod scores were calculated between the markers and a trial map order was produced by applying the seriation algorithm of Buetow and Chakravarti (1987). These orders were used as

---

Figure 2 Representation of genetic maps of maize chromosome 9 from markers typed in three reference panels. Markers are labeled numerically from 1 through 32. The maps are scaled to distance in centimorgans, with the position of loci indicated by number above the lines == (odds >1000:1 in favor of linkage) or ++ (odds >100:1 but <1000:1 in favor of linkage). Completely linked markers are indicated by ',' joining the two loci numbers. Approximately 1-lod-unit confidence intervals for the placement of loci are indicated below each map. Rearrangements of loci under alternative orders with odds within 100:1 of the best-supported order are shown above each map. No rearrangements of completely linked loci have been examined.

initial starting points for application of several iterations of the gene mapping system (GMS) to find alternative orders with improved likelihood. Details of this algorithm, which involves the application of criteria for selecting subsets of orders to be tested, are given in Lathrop et al., (1988a). Efficient methods to evaluate likelihoods of the gene order in backcross and F_2 panels are described in Lathrop and Lalouel (1990).

Figure 2 shows the final orders obtained in the three mapping panels by convergence under the GMS criteria. Also indicated are the alternative orders tested during the construction that have relative odds within 100:1 of the best-supported order. Except where indicated in Figure 2, inversions of adjacent loci produced odds >100:1 against. Approximate 1-lod-unit confidence intervals are given for the placement of each locus with the map, as calculated by the location score method, as described in Lathrop et al. (1988a).

Although the orders of loci common to the NC and MG panels or the NC and 37 panels are the same in the corresponding maps, the recombination estimates and confidence intervals suggest that the genetic distances may be heterogeneous. Therefore, we tested the heterogeneity of the recombination estimated for marker pairs that were common to the two panels. Tests of heterogeneity were significant between the NC and MG panels 253–WX1 and WX1– 293. Assuming different recombination values led to the estimates $\Theta = 0.46$ and 0.20 for 253–WX1 and $\Theta = 0.12$ and 0.26 for WX1–293, respectively, in the NC and MG panels. The likelihood ratio tests of heterogeneity (ratio of the likelihood assuming different recombination rates and the likelihood assuming a common recombination rate) gave $\chi^2_1 = 9.5$ ($p < 0.005$) and $\chi^2_1 = 4.9$ ($p < 0.05$). Although these results suggest that the linkage maps of maize chromosome 9 may vary as a function of the genetic background, they should be interpreted with caution given the number of comparisons made (heterogeneity tests were performed for seven pairs of loci) and the absence of confirmation in a replicate experiment.

LINKAGE MAPS AND TRAIT PHENOTYPES

Linkage studies usually occur in two phases. First, well-spaced markers are selected for characterization in informative pedigrees or experimental crosses to obtain primary localizations of the genes responsible for the trait phenotype. Although this is usually achieved by pairwise linkage analysis between the trait marker loci, multilocus analysis can also be employed, since complete linkage maps of the markers are often available. The first phase of the study ends when significant linkage is found to one or more marker loci,leading to a preliminary assignment of a gene or genes responsible for the trait phenotype.

The goal of the second phase is to determine precise localizations for the trait genes, ideally by placement of genetic markers at 1% recombination distance about the trait loci. Often, this stage requires development of new makers and construction of a high-resolution linkage map of a specific chromosome region. A combination of linkage data from reference families or meioses and others informative for the trait phenotype may be necessary to obtain a linkage map of this resolution.

The location score method (Lathrop et al., 1985) provides one approach to obtain an initial or precise localization of a trait locus. In this approach, a genetic map of the chromosome or chromosome region of interest is constructed from data on a reference pool. The map of marker loci is assumed to be fixed, and likelihoods for the placement of the trait at various positions on the map are calculated by multilocus methods. The location score for a position d on the map is defined as $-2\ \ln[(L(d))/L(\text{unlinked})]$, where L(unlinked) is the likelihood with the trait locus not linked to the markers in the map.

In certain experimental crosses, location scores can be calculated by considering only the two closest markers flanking an interval; this simplified situation has been called “interval mapping” (Lander and Botstein, 1989). When data are obtained from outbred populations, this simplification does not generally apply, and more loci must be accounted for in the calculations.

EXAMPLE: LOCATION SCORE APPROACH APPLIED TO ADENOMATOUS POLYPOSIS COLI

The gene responsible for adenomatous polyposis coli (APC), an autosomal dominant form of cancer in humans, was mapped to chromosome 5q21-22 by linkage studies with the probe c11p11 (D5S71) (Lep-

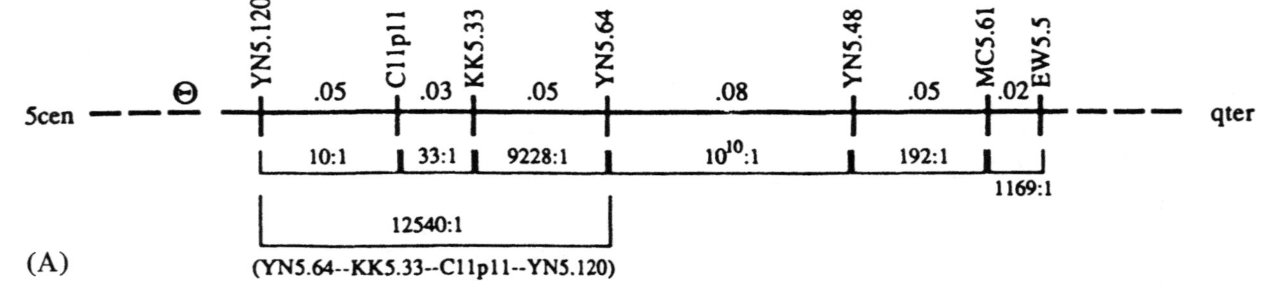

Figure 3 (A) High-resolution linkage map of seven markers in the region of the APC gene, odds against inversion of adjacent loci, and the orientation of a subgroup of loci. (B) Location score curves showing the maximum-likelihood placements of the APC locus with respect to the map of marker loci. The solid line indicates location scores with respect to a subset of loci C11p11, KK5.33, YN5,64, and YN5.48; the dashed line indicates location scores with respect to the subset YN5.64, YN5.48, MC5.61, and EW5.5. Both curves place the maximum-likelihood location of ACP very close to YN5.48 (Reprinted with permission of the American Society of Human Genetics from Nakamura et al., 1988.)

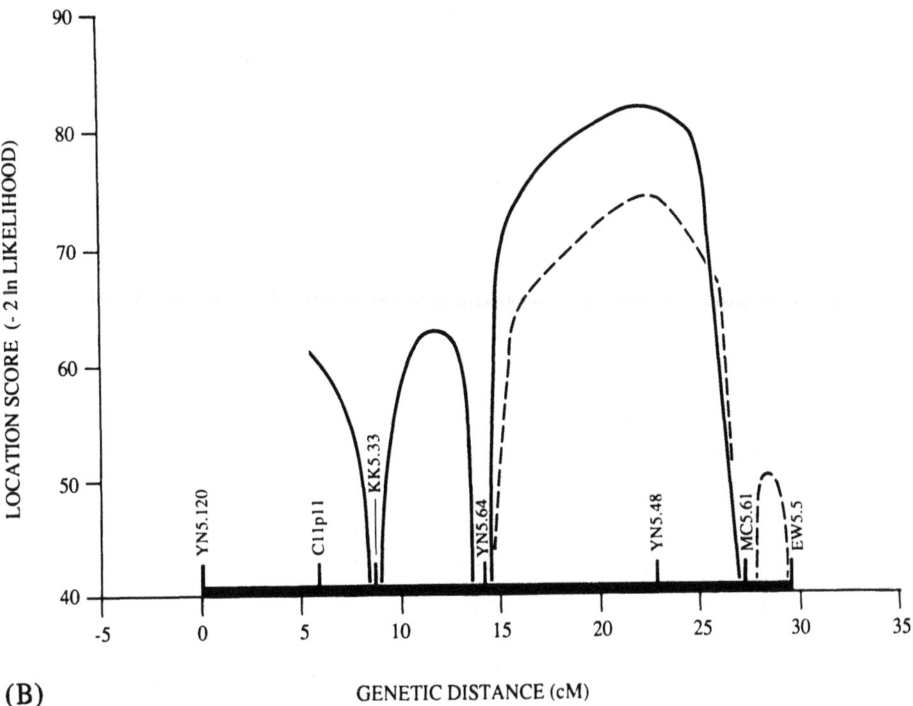

Figure 3 Continued.

pert et al., 1987; Bodner et al., 1987). These studies were motivated by a report of a partial deletion of the long arm of chromosome 5 in a patient with mental retardation (Herra et al., 1986). The combines maximal lod score for the two initial linkage studies was 6.7 at $\Theta = 0$.

To provide more precise evidence on the localization of the APC gene, Nakamura et al. (1988) screened a cosmid library constructed from a somantic cell hybrid containing 40% of the long arm of chromosome 5 as its human component. Several of these could be mapped to the region of APC by close linkage to C11p11 when mapped in 59 reference families (the CEPH panel plus 19 additional, well-characterized three- or two-generation families). A genetic map containing seven marker loci, including C11p11, from this region of chromosome 5 is shown in Figure 3 (a). Except for possible alternative placements for the two markers closest to C11p11, the order of the loci was well-determined on the basis of the likelihood ratio criterion.

Statistical analysis of six large families ascertained for presence of the disease showed that 5 of these markers exhibited lod scores >3 with ACP, while the other 2 exhibited lod scores > 2. To improve knowledge of the placement of the disease gene, location scores were calculated assuming the placement of the markers as shown in Figure 3A. The results, given in Figure 3B gave strong evidence in favor of a placement of the gene responsible for ACP near the marker YN5.48, which was estimated to be 17 cM distal to C11p11.

FURTHER ISSUES IN DATA MANAGEMENT

High-resolution linkage and physical maps of chromosomes on humans and other species will be developed over the next few years, using a variety of techniques, including linkage studies. In humans, ordering of loci in 1-5 cM linkage maps is theoretically possible with the extended CEPH panel, consisting of the 40 original families plus 19 others (Lathrop et al., 1988b), and genotypes for between 5000 and 10,000 markers are expected to be generated on this panel. The creation of efficient means for rapid dissemination will be essential. Many highly polymorphic markers, such as microsatellites (Litt and Luty, 1989; Weber and May, 1989), can be detected by polymerase chain reaction-based methods, and it is likely that these systems will form the basis of many high-resolution linkage maps; the concept of sequence-tagged sites (STSs) (Olson et al., 1989) provides a natural way of linking information on the genetic and physical maps. High-resolution linkage studies of the CEPH panel will also benefit from a data base of recombinant meioses, since the effort in constructing the map may be reduced by characterization only of those individuals that exhibit recombination events that could order closely linked markers.

CONCLUSION

The advent of new DNA technology has led to rapid development of genetic marker data and new techniques for linkage studies. It is likely that the realm of these applications will continue to expand as biological tools improve over the next few years. New challenges for data management and analysis can be expected to result from this expansion.

REFERENCES

Bodmer W, Bailey C, Bodmer J, Bussey H, Ellis A, Gorman P, Lucibello F, Murday V, Rider S, Scambler P, Sheer D, Solomon E, Spurr N. Localization of the gene for familial adenomatous polyposis on chromosome 5. Nature 1987; 328:614-616.

Bonney G, Lathrop GM, Lalouel JM. Combined linkage and segregation analysis using regressive models. Am J Hum Genet 1988; 43:29-37.

Buetow K, Chakravarti A. Multipoint mapping using seriation. Am J Hum Genet 1987; 41:180-201.

Burr B, Burr F, Thompson KH, Albertson MC, Stuber CW. Gene mapping with recombinant inbreds in maize. Genetics 1988; 118:519-526.

Clayton J. A multipoint linkage analysis program for X-linked disorders, with the example of Duchenne muscular dystrophy and seven DNA probes. Hum Genet 1986; 73:68-72.

Dausset J, Cann H, Cohen D, Lathrop GM, Lalouel JM, White R: Centre d'Etude du Polymorphism Humain (CEPH): Collaborative genetic mapping of the human genome. Genomics 1990; 6:575-577.

Donis-Keller H, Green P, Helms C, Cartinhour S, Weiffenback B, and 28 others. A genetic linkage map of the human genome. Cell 1987; 51:319-337.

Everitt BS, Hand DJ. Finite mixture distributions. London: Chapman and Hall, 1981.

Georges M, Lathrop M, Hilbert P, Marcotte A, Schwers A, Swillens S, Vassart G, Hanset R. On the use of DNA fingerprints in mapping the bovine "muscular hypertrophy" gene. Genomics 1990; 6:461-474.

Hanset R, Michaux C. On the genetic determination of muscular hypertrophy in the Belgian white and blue cattle breed. I. Experimental data. Genet Sel Evol 1985; 17:359-368.

Hasstedt SJ. Linkage analysis using mixed, major gene with generalized penetrance or three-locus models. Hum gene mapping 6. Cytogenet Cell Genet 1982; 18:284-285.

Helentjaris T, Slocum M, Wright S, Schaefer A, Nienhuis J. Construction of genetic linkage maps in maize and tomato using restriction fragment length polymorphisms. Theor Appl Genet 1986; 72: 761-769.

Herra L, Kakati S, Gibas L, Pietrzak E, Sandberg A. Brief clinical report: Gardner syndrome in a man with an interstitial deletion of 5q. Am J Med Genet 1986; 25:473-476.

Lalouel JM, Lathrop GM, White R. Construction of human genetic linkage maps. II. Methodological issues. Cold Springs Harbor Symp 1986; 51:39-48.

Lander ES, Botstein D. Mapping mendelian factors underlying quantitative traits using RFLP linkage map. Genetics 1989; 121:185-199.

Lander ES, Green P. Construction of multilocus linkage maps in humans. Proc Natl Acad Sci USA 1987; 84:2363-2367.

Lander ES, Green P, Abrahamson J, Barlow A, Daly MJ, Lincoln SL, Newbers L. MAPMAKER: an interactive computer package for constructing primary genetic linkage maps of experimental populations. Genomics 1987; 1:182-195.

Lange K, Weeks D, Boehnke M. Programs for pedigree analysis: MENDEL, FISHER, and dGENE. Genet Epidemiol 1988; 5:471-462.

Lathrop GM, Lalouel JM. Simple calculation of lod-scores on small computers. Am J Hum Genet 1984; 36: 460-465.

Lathrop GM, Lalouel JM. Efficient computations in multilocus linkage analysis. Am J Hum Genet 1988; 42:498-505.

Lathrop M, Lalouel JM. Statistical methods for linkage analysis 1990 (in press).

Lathrop M, Nakamura Y. Mapping the human genome. In: Karan J, Chao L, Wan G, eds. Methods in nucleic acids research. Boca Raton, FL, CRC Press, 1991:157-180.

Lathrop GM, Lalouel JM, Julier C, Ott J. Strategies for multilocus linkage analysis in humans. Proc Natl Acad Sci USA 1984; 881:3443-3446.

Lathrop GM, Lalouel JM, Julier C, Ott J. Multilocus linkage analysis in humans: detection of linkage and estimation of recombination. Am J Hum Genet 1985; 37:482-498.

Lathrop GM, Lalouel JM, White RL. Calculation of human linkage maps: likelihood calculations for multilocus linkage analysis. Genet Epidemiol 1986; 3:39-52.

Lathrop GM, Nakamura Y, Cartwright P, O'Connell P, Leppert M, Jones C, Tateishi H, Bragg T, Lalouel JM, White R. A primary genetic map of markers for human chromosome 10. Genomics 1988a; 2:157-164.

Lathrop GM, Lalouel JM, White R. The number of meioses needed to resolve gene order in a 1% linkage map. Human gene mapping 10. Cytogenet Cell Genet 1988b; 46:643.

Leppert M, Dobbs M, Scambler P, O'Connell P, Nakamura Y, Stauffer D, Woodward S, Burt R, Hughes J, Gardner E, Lathrop M, Wasmuth J, Lalouel JM, White R. The gene for familial polyposis coli maps to the long arm of chromosome 5. Science 1987; 238:1411-1413.

Litt M, Luty JA. A hypervariable microsatellite revealed by in vitro amplification of a dinucleotide repeat within the cardiac muscle action gene. Am J Hum Genet 1989; 44:397-401.

Monaco A, Neve R, Colletti-Feener C, Bertelson C, Kurnit D, Kunkel L. Isolation of candidate cDNAs for portions of the Duchenne muscular dystrophy gene. Nature 1986; 323:646-650.

Morton NE. Sequential tests for the detection of linkage. Am J Hum Genet 1955; 7:277-318.

Nakamura Y, Lathrop GM, Leppert M, Dobbs M, Wasmuth J, Wolff E, Carlson M, Fujimotot E, Krapcho K, Woodward S, Hughes J, Burt R, Gardner E, Lalouel JM, White R. Localization of the genetic defect in familial adenomatous polyposis within a small region of chromosome 5. Am J Hum Genet 1988; 43:638-644.

Nakamura Y, Lathrop GM, O'Connell P, Leppert M, Lalouel JM, White R. A small region of human chromosome 14 shows a high frequency of recombination. Genomics 1989; 4:76-81.

Olson M, Hood L, Cantor C, Botstein D. A common language for physical mapping of the human genome. Science 1989; 245:1434-1435.

Ott J. Estimation of the recombination fraction in human pedigrees: efficient computation of the likelihood for human linkage studies. Am J. Genet 1974; 26:588.

Ott J: Analysis of human genetic linkage. Baltimore: Johns Hopkins University Press, 1985.

Ott J: Computer-simulation methods in human linkage analysis. Proc Natl Acad Sci USA 1989; 86:4175-4178.

Plougman LM, Boehnke M. Estimating the power of a proposed linkage study for a complex genetic trait. Am J Hum Genet 1989; 44: 513-527.

Riordan JR, Rommens JM, Kerem B, Alon N, Rozmahel R, Grzelczk Z, Zielenski J, Lok S, Plavsic N, Chou JL, Drumm M, Iannuzzi MC, Collins FS, Tsui LC. Identification of the cystic fibrosis gene: cloning and characterization of complementary DNA. Identification of the cystic fibrosis gene: cloning and characterization of complementary DNA. Science 1989; 245:1066-1072.

Smith CAB: Some comments on the statistical methods used in linkage investigations. Am J Hum Genet 1959; 11:289-304.

Taylor B: In: Lyon M, Searle A, eds. Genetic variants and strains of the laboratory mouse, 2nd ed. New York: Oxford University press, 1989: 773-796.

Weber R, Leppert M, Bishop T, Barker D, Berkowitz J, Brown C, Callahan P, Holm T, Jerominski L. Construction of linkage maps with DNA markers for human chromosomes. Nature 1985; 313:101.

White R, Lalouel JM. Investigations of genetic linkage in human families. In: Harris H, Hirschhorn K, eds. Advances in human genetics, Vol 16. New York, Plenum Press, 1987:121–228.

APPLICATIONS OF GENE MAPPING

10

Genomic Genetics and Plant Genetic Improvement

Jacques S. Beckmann

The Volcani Center
Bet Dagan, Israel

INTRODUCTION

The ability to generate complete genetic maps includes the capacity to evaluate the entire genome (hence the term "genomic genetics"), to dissect complex genetic traits into their individual mendelian entities, and to channel all this information into breeding programs. This is what renders marker-based methodologies so powerful. The progress to date in plant genetics will be reviewed and the limitations outlined.

Mapping techniques require still further refinement. For some plants, polymorphism is abundant and easy to uncover. For others, as well as for specific narrow crosses, the opposite situation prevails; informative markers are also required in these situations. Possible strategies to uncover such markers will be discussed.

The search for additional useful markers can be complemented by the design of more efficient mapping schemes. One design, which has remarkable qualities both for the identification and evaluation of single

---

Contribution from the Agricultural Research Organization, The Volcani Center, Bet Dagan, Israel, No. 2832-E, 1989 series.

trait loci of economic importance (ETL) and for precise ordering of loci on the chromosome map will be presented. This will enable markers to be placed within realistic "chromosome" walking distances of the ETL, allowing for the molecular cloning of these loci by reverse genetics.

Thus genomic markers, besides serving as specific "tags" to monitor for the presence of ETL in breeding programs, can also serve as reference points toward the cloning of genes of interest. In this way, genomic genetics can provide a solution to two seemingly unrelated problems—the molecular cloning of genes of agricultural interest and the more effective utilization, by classical breeding techniques, of existing genetic variation within a species—and become an intrinsic component of modern breeding methodologies.

Genetic improvement is an area in which new technologies are often slowly assimilated. The latest newcomer in the breeder's armamentarium is recombinant DNA technology, which provides greatly increased genetic resolution. Indeed, it is now possible to uncover large numbers of genetic polymorphisms, to locate them accurately on the genetic map, and to use them as markers for the evaluation and utilization of the genetic basis for observed phenotypic variability. This is done by uncovering genetic variation directly at the DNA level, in the form of genomic markers. The capacity to relate directly to genomic variation rather than having to infer genotypes from phenotypic observations (the classical mendelian paradigm) led to proposal of the term "genomic genetics" (Beckmann, 1988).

DNA sequence variation is usually equated with restriction fragment length polymorphism (RFLP). The latter, however, is only one of a number of means that can be utilized to uncover genomic variation (we might say, by analogy with computers, that we have already seen several generations of genomic markers). Our purpose is not to review this evolution. The emphasis lies instead on markers in general, rather than on the exact type of markers utilized, and on the fact that genomic markers (whose convenient and attractive properties were reviewed in Beckmann and Soller, 1986) can be used by breeders to deal with the problems posed by traits under the control of a complex of polygenic genetic factors—the quantitative trait loci. These principles will be illustrated with some examples taken from plant studies.

MARKER-BASED METHODOLOGIES

The detailed interrelationships between markers or markers and trait loci need to be determined. But even before this goal is reached, the availability of a battery of markers, per se, lends itself to some immediate and important potential applications (Table 1), such as the assignment of breeding lines to heterotic groups for the prediction of single cross performances (e.g., Kahler and Wehrhahn, 1986; Lee et al., 1989) or the demonstration of the distinctiveness of newly released varieties, a central issue in varietal protection. In fact, genomic markers are the

Table 1 Applications of Marker-Based Methodologies

- In conventional breeding
  - Short-range applications
    - Parentage identification
    - Protection of breeder's rights
    - Assignment of inbred lines to heterotic groups
    - Line and hybrid purity testing
  - Long-range applications
    - Analytical path
      - Mapping and evaluation of effects of single ETL
      - Exploration of homologous loci in other species or genera
      - Germplasm evaluation (diversity, classification, and phylogeny)
    - Synthetic path
      - Introgression of desired traits among cultivars or from resource strain to cultivar
      - Preselection of superior genotypes
      - Prediction of expected phenotypes
- In transgenic breeding
  - Precise mapping, a first step toward the cloning of ETL and their manipulation through genetic engineering techniques (transgenics)

most adequate tool available today to estimate minimum genetic distance between lines that can often be morphologically and agronomically similar. These markers are, therefore, gradually becoming recognized by courts in litigations in plant breeding over misappropriated proprietary germplasm. Genomic markers have indeed the necessary attributes required for these tasks: they provide an adequate sampling of the genome, they are overall the most sensitive probes available for polymorphisms, they are both genetically and environmentally stable, they act in a codominant manner, and they can reproducibly be measured. In addition to the utilization of the existing genetic variation, genomic markers can also be generated do novo: a foreign segment of DNA, once it is stably inserted into the genome, constitutes an easily scorable tag (e.g., Beckmann and Bar-Joseph, 1986).

But the foremost application of genomic markers is elucidation of the historic, genetic, and functional architecture of a genome. It is their enhanced genetic resolution power which is at the basis of "marker-based methodologies." The latter will enable the analysis of related genomes, the establishment of taxonomical inferences, such as chromosomal duplications or alloploidy formation (see, for instance, Song et al., 1988), the tracing of the evolution of specific cultivars and thus, the verification of the validity of pedigree data. In this manner markers can provide important clues on germplasm diversity in the relatives of domesticated plants. More importantly, marker-based methodologies should make it possible to examine intragenomic relationships, namely, the analysis of the genome into its individual functional components, the mendelian entities, leading to a precise evaluation and understanding of their effects (Beckmann and Soller, 1986) and, eventually, to the molecular cloning of the loci involved and to transgenic manipulations (Table 1).

When considering marker-based methodologies, it is opportune to remember that there are two facets. The first—the subject of this discussion—is analytical: the identification and evaluation of the individual genetic factors—the economic trait loci (ETL)—governing both simple and polygenic traits, allowing for exploration of the entire species gene pool for a search for beneficial alleles. The second is synthetic: integration of all the information gathered on relative breeding values of the characterized loci, allowing predictions to be made as to the overall phenotype to be expected upon combining various alleles and providing the necessary markers for the controlled and efficient introgression of defined chromosomal segments into desired genetic backgrounds. This can replace the less accurate and often difficult scoring of the trait itself (phenotypic selection). Progeny bearing the selected markers (genomic selection) will be enriched for the desired trait; this will be equivalent to selection with a heritability of close to 1.0. Even for easily scorable traits, marker-assisted selection can, in special circumstances, provide an attractive alternative; for instance, wherever there are constraints on the introduction of undesirable pathogens, selection for resistant genotypes could still be pursued without colonization.

TOWARD SATURATED MAPS

The rationale of marker-based methodologies lies in the capacity to distinguish allelic pairs of markers and to assess the phenotypic value of the associated chromosomal segments. By following marker alleles, it is possible to recognize which of the alternative chromosomal regions defined by a given marker was inherited by each individual and, by suitable experimental design, to determine the polygene allele(s) coupled with each of the marker alleles, leading to the identification of genes affecting the economic traits considered (Beckmann and Soller, 1986; see also Edwards et al., 1987; Grant et al., 1988; Kahler and Wehrhahn, 1986; Landry et al., 1987; Martin et al., 1989; Nienhuis et al., 1987, Osborn et al., 1987; Paterson et al., 1988; Tanksley et al., 1982; Tanksley and Hewitt, 1988; Weller et al., 1988).

The looser the linkage between marker and gene, the more difficult the identification. Hence, the necessity for a large number of markers, densely saturating the entire genome. Saturation, however, is a relative term. It depends on the actual application—classical or molecular genetics. A molecular cloner would find it hard to settle with a map consisting of but a couple of markers per chromosome, say every 20 cM, whereas a breeder might be quite satisfied with such a situation. Monte Carlo simulations show that in order to achieve an 80% genome coverage at a 20-cM maximum spacing between markers, approximately 90 randomly distributed markers are required per 1000 cM (Kashi et al., 1986). Thus, the number of markers required for adequate genome coverage (from a breeder's point of view) is well within reach, as evidenced

Table 2 Molecular Genomic Markers Already Reported in Plants

Arabidopsis	>100
Barley	>30
Brassica	>250
Lettuce	>120
Lentil	>20
Maize	>900
Pepper	>80
Potato	>300
Rice	>135
Soybean	>115
Tomato	>400
Wheat	>150

by the rapid progress made in this area (Table 2; see also Bonierbale et al., 1988; Chang et al., 1988; Gebhardt et al., 1989; Havey and Muehlbauer, 1989; Helentjaris and Burr, 1989; Helentjaris et al., 1986; Landry et al., 1987; McCouch et al., 1988; Tanksley et al., 1988a,b).

It might be informative to briefly review the means utilized by plant geneticists to generate these maps. Basically, the considerations can be divided into three: the nature of the cross examined, the genetic shortcuts, and the probes utilized. With selfers, interspecies crosses were used to optimize for genetic variation (e.g., Bonierbale et al., 1988; Tanksley et al., 1988a,b). With other plants, such as maize, this precaution was not necessary as for almost all probes considered, an enzyme combination could be found that would render it informative (Evola et al., 1986; Burr et al., 1988).

Map assignments are typically done on a segregating F_2 or backcross population, with up to as many as 100 offspring. Final chromosomal assignment often depends on the identification of linkage to well-known landmarks (such as isozyme markers or other previously mapped loci), allowing for the coordination of the RFLP map to the conventional map. In some species advantage can be taken for chromosomal assignments of the existence of monosomic (2n–1), trisomic (2n+1), of other aneuploid or chromosomal addition lines, or even of well-known translocations, such as the maize B–A translocations (e.g., Evola et al., 1986; Helentjaris et al., 1988; Sharp et al., 1989; Weber and Helentjaris, 1989). Translocations and other chromosomal rearrangements have the added advantage of enabling the alignment of the physical and genetic maps.

It should be stressed that in these mapping schemes each plant (or population) cannot be reproduced sexually and is in fact unique. Hence, unless every individual is being propagated (e.g., by vegetable cloning or as an individual F_3 families), or its DNA conserved, analogous populations will have to be generated anew in each study. A significant improvement was recently introduced by Burr et al., (1988), who demonstrated the usefulness of recombinant inbred strains (RIL) for RFLP mapping. This approach, when applicable, has important advantages: The same segregating RILs can be replicated at will and shared by many geneticists working separately, each adding, in a cumulative manner, new markers onto the map. All this can be done practically, as illustrated by their work, with a minimal number of Southern blotting experiments.

Almost all RFLP probes utilized so far were single-copy DNA probes, with a few exceptions corresponding to multigene families. Some probes were obtained by taking advantage of the fact that most genes are transcribed from single-copy DNA (e.g. Bonierbale et al., 1988; Burr et al., 1983; Rom, 1982; Tanksley et al., 1988a,b) and that cDNA libraries are easy to obtain. Alternatively, genomic libraries enriched for singly-copy DNA were obtained upon digestion of the DNA with methylation-sensitive restriction enzyme, such as PstI. Most of the small fragments present in such a maize genomic library proved to be highly enriched for single copy sequences (Burr et al., 1988; Helentjaris et al., 1988). Similar genomic libraries yielded informative probes in a number of additional plant studies (e.g., Bonierbale et al., 1988; Tanksley et al., 1988a,b). Moreover, in some cases, heterologous probes were also utilized. For instance, Bonierbale et al. (1988) reported that of the cDNA and genomic tomato probes examined, all but two failed to hybridize to potato. Similar results were obtained in the study on species from the *Triticeae* tribe (Sharp et al., 1989). It should be noted that despite the number of studies using cDNA or random genomic probes, no general conclusion can be drawn regarding the optimal source of RFLP probe or the most informative enzyme-probe combination. This seems to be a case-by case issue.

Table 2 thus testifies to the rapid progress in construing genetic maps. In the not-too-distant future, the detailed and, for all practical breeding purposes, saturated maps of most agronomically important (plant or animal) species will be available. Still, it should be remembered that the ease with which DNA markers can be uncovered among different plant species varies widely. For some, like maize and *Brassica,* almost all probes screened turn out to be informative with one or the other enzyme tested. In addition, the loci identified often show multiple isoalleles. For other species, such as tomatoes and soybean, the assembly of a large battery of informative genetic markers is much more difficult. Furthermore, the extent to which informative probes can be gathered varies also according to the genetic distance separating the lines to be examined: the closer the lines, the more difficult it becomes. There are thus two different basic issues here: the generation of a saturated genetic "species" map and the identification of highly informative markers that will be applicable to both wide and narrow crosses.

At present even the best available species maps still contain unmarked zones. Covering the last dark regions of the map, by continued accumulation of random markers, may require immense effort. A number of molecular genetic tools, such as pulsed-field gel electrophoresis (Smith and Cantor, 1987), megacloning in yeast (Burke et al., 1987), or jumping libraries (Poustka and Lehrach, 1986), can be used for this purpose. As an additional help in this endeavor, it was proposed (Beckmann and Soller, 1989b) to take advantage of the fact that genome evolution seems to have occurred by reshuffling of large chromosomal blocks during speciation, as was observed for mammalian genomes (Nadeau, 1989; Womack and Moll, 1986). This implies that genes that are near each other in one species have a high likelihood of having maintained this relationship in other species as well, a phenomenon termed syntenic homology. The latter is also likely to apply to the plant genomes: the homosequential gene order reported for tomatoes and potatoes (Bonierbale et al., 1988) and the extensive gene duplication and gene order conservation of the maize (Helentjaris et al., 1988) and *Bras-*

sica genomes (Slocum, 1989) can all be taken as evidence for synteny in plants. Conservation of syntenic relationships, in some instances, can apply to large chromosomal segments, as in the tomato-potato comparison, whereas in other cases (e.g., Tanksley et al., 1988a) it may involve much smaller segments (say of 10–20 cM).

Expanding the concept of synteny to the entire genome implies that it might suffice to have a good "reference" map for one species, in order to have, at first approximation, the "local" map of most other cultivated plants. Thus, in order to render mapping cumulative between plant species, it is important to start using known landmarks; i.e., nonanonymous DNA probes, such as cDNA probes, demonstrate the generality of syntenic homology and build in this manner a reference plant genome map (Beckmann and Soller, 1989b). Comparative mapping would then enable extrapolation from one species to another, leading to generation of adequately saturated genetic maps. Its small genome (15 times that of *Escherichia coli*), near absence of interspersed repetitive DNA, and high genetic prolificity are among the unique attributes that make *Arabidopsis thaliana* an attractive plant model for both classical and molecular genetic studies (Meyerowitz, 1989). Thus, it is no surprise that *A. thaliana* is the first plant for which both a contiguous map of overlapping clones representing the entire genome (Haughe et al., 1989) and an RFLP map (Chang et al., 1988) are near completion. This also makes it an optimal candidate for such a reference map. Comparative mapping, in reference to *A. thaliana*, will be particularly useful for the molecular cloning of loci affecting traits of economic interest mapped to the interval between a pair of marker loci. In this case, it will be possible first to search for the putative locus among *A. thaliana* clones corresponding to the chromosomal segments defined by the flanking markers. Following identification of potential loci in *A. thaliana*, the corresponding clones can be used as probes to retrieve the homologous genes in the species of interest.

Later Generations of Genomic Markers

The necessity to provide discriminative markers for those species showing little polymorphism, as well as for intervarietal crosses, implies the need for yet a higher genetic resolution. Gebhardt et al. (1989) described Southern blots of sequencing gels—upon digestion of the DNA with restriction enzymes having a 4-bp specificity—to uncover extensive polymorphism in potato lines. Using this system, 90% of the probes tested uncovered useful variation in potatoes, several of these markers being oligoallelic. Provided that the suitability of this approach to more difficult situations is confirmed, this could represent a significant methodological improvement.

Since a genome does not evolve at a uniform rate, another promising approach would be to identify the faster-evolving loci and then monitor variation at these sites. This is why the discovery of DNA probes detecting hypervariable (HVR) loci represents a great leap forward in genomic genetics (the second marker generation, sic!). Variation at these loci is generally due to the presence at multiple sites along the genome of some core minisatellite sequence(s) (from 15 to 60 bp long), which is tandemly repeated, with polymorphism being the result of a change in copy number. The presence HVRs showing a common core sequence was first demonstrated in humans (Jeffreys et al., 1985; Nakamura et al., 1987; Vassart et al., 1987) and later extended to animals (e.g., Vassart et al., 1987). These probes tend to be extremely informative, with multiple alleles being identified at each locus (for a review see Jeffreys, 1987). Consequently, hybridization of genomic DNA with one of these probes generates highly genotype-specific fingerprints.

A second family of small reiterated sequences, termed "microsatellites" (Litt and Luty, 1989), was also found to generate mendelianly inherited DNA sequence variation. The latter could be visualized as delicate-length polymorphism (Litt and Luty, 1989; Weber and May, 1989) by polymerase chain reaction (PCR) (Saiki et al., 1988) in sequencing gels or as changes of a grosser nature by RFLP methodologies (e.g., Kashi et al., 1990b).

Whereas the occurrence of mini-and microsatellite sequences in the animal kingdom is firmly established, preliminary examinations of a number of plant species (Dallas, 1988; Nybom et al., 1990; Rogstadt et al., 1988; Ryskov et al., 1988) with hypervariable minisatellite DNA probes also yielded variable DNA fingerprints. These results support the contentions of (1) ubiquity of many of these sequences and (2) a high level of polymorphisms associated with the sites uncovered by these probes. This raises the exciting possibility that minisatellites probes could be of wide applicability in many plant genetic programs. In contrast, the suitability of microsatellites as generators of genomic polymorphisms in plants still remains to be demonstrated.

One particular application area of HVR probes merits mentioning, namely, their utilization in intercultivar comparisons. Indeed, many breeding programs involve the generation of superior varieties by crossing existing top varieties. The problem in so doing is that these varieties are agronomically and morphologically fairly similar (and often near-isogenic), so that it may be hard to tell them apart with all markers available. Provided that the observations made in mammals can be extrapolated to plants, the great plasticity of HVR loci makes them at present the optimal means for generating cultivar-specific fingerprints, allowing for the discrimination between closely related lines. Whereas this possibility is still in the realm of a hypothesis (which is currently being examined), we are extremely eager to see the results of intervarietal comparisons with these probes.

These are, however, important constraints in the utilization of HVR probes. Whereas the genetic analysis, i.e., the establishment of allelic relationships in inbred lines, is essentially feasible, the examination of the segregation of HVR loci in outcrosses is generally more problematic. This is due to the fact that it is hard to relate one particular band in the gel to its allelic counterparts in each individual. There is one notable exception. Consider a cross between two heterozygous individuals (e.g., two vegetatively propagated trees). Its F_1 progeny could be examined and interpreted in terms of the segregation of the four alternative alleles at each locus and linkage relationships established.

These constraints can be solved, however, by concentrating on each of these loci—one at a time. This can be done by focusing on the single-copy sequences flanking the tandemly repeated cores at each locus and utilizing them as RFLP probes. These variable number of tandem repeats probes (VNTR; Nakamura et al., 1987) can then be utilized as highly polyallelic, locus-specific DNA probes, with the potential promise of being optimally discriminative in most genetic programs. Another promising way to visualize locus-specific variation is by measuring the subtle changes associated with microsatellite sequences, of which $(TG)_n$ seem to be the most abundant representative in vertebrates (e.g.,

Hamada et al., 1982). Microsatellites offer several important advantages over minisatellites. Virtually all poly-TG containing loci studies in humans appear to show delicate length variation (i.e., confined to few base pairs), the markers so uncovered bearing a high information content (Litt and Luty, 1989; Weber and May, 1989). Moreover, while it appears that minisatellites may be confined to some chromosome sites only (Royle et al., 1988), poly-TG islets are distributed widely throughout the genome, with little tendency to cluster (Hamada et al., 1982; Kashi et al., 1990a; Litt and Luty, 1989). If uniformly spaced, these microsatellites would fall, on the average, every 50–100 kb. Their abundance, informativeness, and apparent wide genomic dispersion imply that markers based on length variation of poly-TG stretches might be found within close walking distances (i.e., a fraction of a centimorgan) of any locus of interest. This would provide the capacity to identify highly informative bracketing markers or haplotypes and would offer interesting new perspectives.

To sum up this discussion, a long distance has been covered since the first RFLP demonstration (Grodzicker et al., 1974) and the proposal to use RFLPs as markers in human genetics (Botstein et al., 1980) or in genetic improvement programs of plants (Beckmann and Soller, 1983; Burr et al., 1983; Rom, 1982; Soller and Beckmann 1983) and animals (Soller and Beckmann, 1982, 1983). Near-complete genome coverage is already a fact for a number of species (Table 2). In the foreseeable future, informative markers will be available also for intervarietal discrimination. Additional markers are constantly accumulating. In parallel, more powerful or less laborious techniques to uncover new types of genomic markers are steadily being introduced (e.g., Beckmann, 1988). The introduction of the PCR (Saiki et al., 1988) in conjunction with the constantly increasing DNA sequence data represents a milestone in this endeavor. What will be the nature of tomorrow's markers? No one knows, other than that they too will be DNA sequence markers, and that they will bestow a hitherto inaccessible fine-detail genetic resolution. Adequate strategies will have to be devised to take advantage of this enhanced power.

CURRENT MAPPING STRATEGIES AND THEIR LIMITS

Most ETL mapping strategies start with the recognition of two parental genotypes, generally inbred lines homozygous for alternative alleles at the loci of interest, or an inbred line and a selfing landrace, followed by the generation of hybrid (F_1) progeny. The classical schemes continue by either selfing the F_1 or backcrossing it to one of the parental lines to generate a segregating F_2 or BC population. The latter is then scored for performance and markers (e.g., Edwards et al., 1987; Kahler and Wehrhahn, 1986; Tanksley et al., 1982; Weller et al., 1988), in order to identify marker-ETL relationships.

The main difficulty with this type of assay stems from the elusive quality of the phenotypic effects of the genes under scrutiny. Because numerous genetic and environmental factors affect overall phenotype, a large number of individuals needs to be scored, both for performance and for marker genotypes, in order to identify an association between marker and ETL. Initial estimations indicated that an F_2 population of approximately 2000 individuals would be required to uncover genes of an effect of 2-3% of the mean (Soller et al., 1976). Most experiments performed to date, however, involved smaller populations.

Current State of Marker-ETL Linkage Studies

A limited number of detailed investigations trying to establish linkage between markers and quantitative trait loci have been reported so far. This reflects (1) the paucity of markers that was, until recently, available, and (2) the time required to complete these studies. Most of them focused on tomato and maize, using morphological and isozyme markers and, more recently, RFLPs, albeit a number of studies on other plants can also be found.

It is impossible to do justice to all who contributed in this area, so that a biased and partial listing of landmarks will be briefly discussed. One of the early studies is that of Vallejos and Tanksley (1983), who examined cold tolerance in an interspecific backcross of tomato. Later, Weller et al. (1988) examined about 1700 individual plants of a segregating F_2 population, resulting from an interspecific tomato cross, with respect to 18 quantitative traits. Major genes affecting the quantitative traits were not found, but a substantial fraction of the marker-trait combinations examined showed significant quantitative effects associated with the genetic markers. Extensive investigation for identifying and locating loci affecting grain yield or other complex yield-related traits were also performed in two large F_2 populations (approximately 1800 plants each) in maize (Edwards et al., 1987; Stuber et al., 1987). The two populations were segregating for about 20 marker loci, distributed on 9 of the 10 maize chromosomes, bringing close to half the genome within 20 cM of one of the markers. Despite the fact that the markers used were few and far apart, a number of significant associations could be uncov-

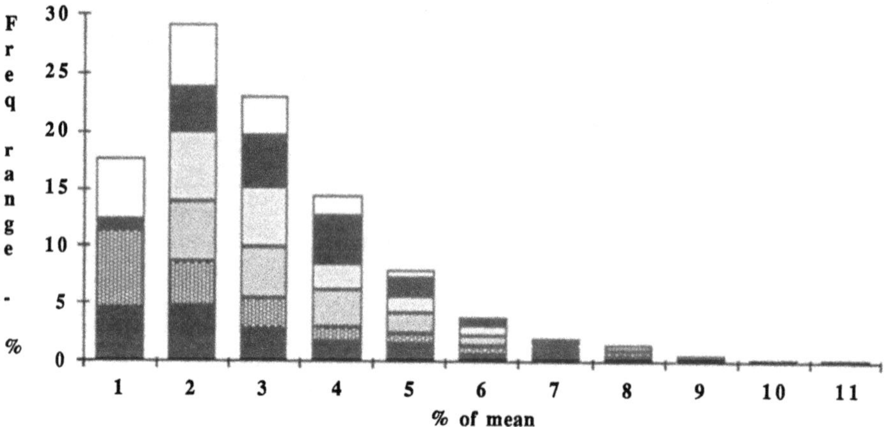

Figure 1 Magnitude of gene effect is given as a function of percent of mean value. When necessary, published data were transformed as follows: from R^2 values to the phenotypic standard deviate, by $2d/\sigma = \sqrt{(8R^2)}$ from $2d/\sigma$ to % of mean by assuming σ to be approximately 10% of the mean. From bottom to top, data are taken from the CMT and COTX crosses from Edwards et al. (1987), additional data from the same crosses from Stuber et al. (1987), and from the crosses of Weller et al. (1988) and Grant et al. (1988). For Grant et al. (1988), percent of mean was calculated directly from the data, by assuming mean plant and ear height to be 200 and 90 cm, respectively. (Reproduced from Beckmann and Soller, 1989b.)

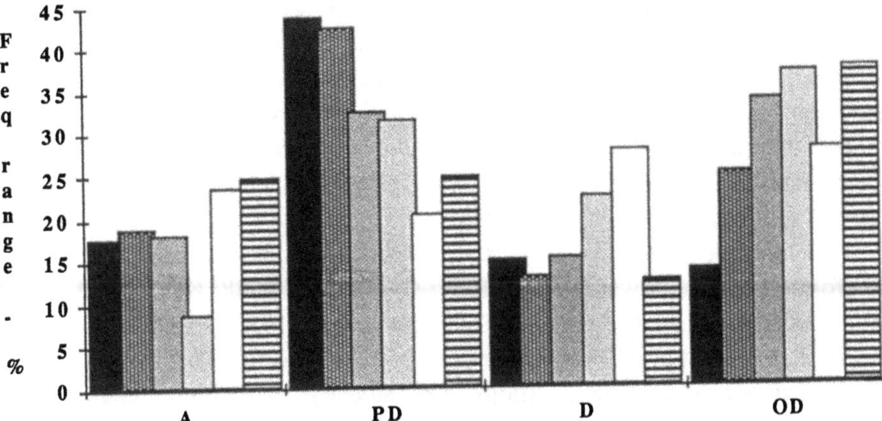

Figure 2 Distribution of marker-linked dominance effects into additive (A: from 0 to 0.2), partial dominant (PD: from 0.21 to 0.80), dominant (D: from 0.81 to 1.20), and overdominant (OD: >1.20) effects. Data, from left to right, are for the CMT and COTX crosses from Edwards et al. (1987), additional data from the same crosses from Stuber et al. (1987), and for the crosses of Weller et al. (1988) and Kahler and Wehrhahn (1986). (Reproduced from Beckmann and Soller, 1989b.)

ered in all these studies, allowing inferences to be made regarding the nature, sign, and magnitude of the effects. These are summarized in Figures 1 and 2.

Osborn et al. (1987) were the first to report an association between RFLPs and quantitative trait loci, specifically, genes controlling soluble solid contents in tomato fruits. Since then, these studies have been extended by Tanksley and Hewitt (1988) and Paterson et al. (1988). Linkage between RFLPs and loci controlling in tomato insect resistance (Nienhuis et al., 1987), water use efficiency (Martin et al., 1989), and fruit mass or fruit pH (Paterson et al., 1988) was also demonstrated. The latter study, based on the analysis of 237 backcross plants, used 70 equidispersed genetic markers, bringing an estimated 95% of the genome within detectable distance from the markers. Such genome coverage also enabled the percentage of recurrent parent genotype in the backcross progeny (which varied from 58.95 to 90.3%, with a mean of 75%) to be measured. This study indicates elegantly how markers could be utilized, provided adequate genome coverage, to take advantage of (or counter—depending on the case) the segregation distortion from the expected mendelian inheritance (meaning that some parental alleles might be either under- or overrepresented in the offspring). This could be essential to speed up selection processes.

Several important general observations can be made from the studies performed so far (Edwards et al., 1987; Grant et al., 1988; Kahler and Wehrhahn, 1986; Martin et al., 1989; Nienhuis et al., 1987 Osborn et al., 1987; Paterson et al., 1988; Stuber et al., 1987; Tanksley and Hewitt, 1988; Tanksley et al., 1982; Vallejos and Tanksley, 1983; Weller et al., 1988). Despite the somewhat small number of markers utilized relative to the number required for genome saturation, a significant number of marker-ETL associations have been uncovered for all species examined (up to 60% of the number of pairwise combinations, in some studies). This number increases disproportionately as either more markers or more traits are scored, since (1) the same ETL may be uncovered independently by several linked markers, (2) the same locus may expert pleiotropic effects on several traits, and (3) a number of measurements of the phenotype may all be different expressions of the same trait. All these reasons combine to inflate the number of potential "ETL" uncovered. For example, in the study by Paterson et al. (1988), different "independent" traits were mapped to the same genetic region. More detailed studies of this nature will eventually determine whether the same ETL affects several traits, or whether these regions contain clusters of loosely linked ETL.

An important aspect of these mapping studies is that they also enable "cryptic" genetic variation to be uncovered, i.e., the identification of beneficial alleles that are otherwise hidden in a sea of deleterious alleles. Weller et al. (1988), for instance, reported that 24% of the total number of marker-associated effects detected were in opposite sign to the overall difference between parental strains, with most traits examined showing at least one significant cryptic effect. This ability to detect cryptic variation can be important in the exploration of native germplasms (Soller and Beckmann, 1988).

Estimates of the magnitude of the effects associated with ETL (i.e., the difference in value between the means of alternative homozygous classes, divided by the means of the population) can also be derived from these experiments (Edwards et al., 1987; Grant et al., 1988; Kahler and Wehrhahn, 1986; Nienhuis et al., 1987; Osborn et al., 1987; Paterson et al., 1988; Stuber et al., 1987; Tanksley and Hewitt, 1988; Tanksley et al., 1982; Vallejos and Tanksley, 1983; Weller et al., 1988). Figure 1 shows that for most traits, in the two species that have been examined in detail (maize and tomato), the effects are approximately 2-4% of the mean. Owing to the recombination between marker and ETL, this value is a slight underestimate of true gene effect. Such a coincidence is satisfying, considering that different traits, species and types of crosses (a cross between two inbred lines or between a Landrace and an inbred line) are involved. Other studies examined only a small number of loci and traits or used a backcross design, and hence it is not possible to integrate results from these studies with the values reported above. Most of them, however, fall near or within the range defined, except for some tomato fruit characteristics where effects above 40% of the mean were obtained by Paterson et al. (1988).

Furthermore, utilization of an F_2 scheme enables dominance relationships at the identified ETL to be estimated as well. To this end, the mean phenotypic values of the three marker genotypes are compared. If the heterozygote falls between the two homozygous genotypes, the trait is said to be codominant or additive. If the heterozygote's value is similar to that of one of the homozygous classes, the trait shows dominance (it can also show partial dominance or overdominance). Figure 2 summarizes the distribution of dominance ratios obtained in a number of studies. The remarkable aspect is the very high proportion of loci showing dominance (PD-D-OD) compared with the usual biometrical assumption of lack of dominance. This is particularly striking in view of the fact that dominance estimates are powerfully attenuated according to the recombination distance separating the marker locus and ETL (Soller and Beckmann, 1983) and that these distances, in the reported experiments, may often have been large. These results were also found in the tomato cross examined by Weller et al. (1988), suggesting that they are not specific to maize.

In contrast to these results, gene-gene interaction (epistatic effects) seems to be a minor component of genetic variation. Few epistatic interactions were observed at the 1% significance level (Nienhuis et al., 1987), and those that were observed at the 5% level are no more numerous than expected by chance alone. This may come as a surprise considering the type of crosses made [e.g., a wide interspecies cross, in the study of Weller et al. (1988), since we might have expected coadapted groups of genes to have emerged during evolution]. This result may reflect the fact that the epistatic contribution to genetic variance is small or else that epistasis is also diminished by recombination events, and hence escaped detection. In support of epistasis, Tanksley and Hewitt (1988) found that the same *Lycopersicon chmielewskii* fragments that are introgressed in different *Lycopersicon esculentum* backgrounds manifested large variation in genetic effects.

REFINEMENTS OF MAPPING STRATEGIES

Designs for Reducing the Number of Individuals Scored for Markers

The design of efficient mapping strategies enabling the number of individuals that must be scored for markers for given power to be reduced, at the expense of evaluating larger numbers with respect to the economic traits of interest, is discussed elsewhere in this book. Basically these strategies fall into two categories, designated "trait-based" designs (Lander and Botstein, 1989; Lebowitz et al., 1986; Soller and Beckmann, 1983) and "replicated progenies." In the latter, a population of lines (such as vegetable clones, individual F_3 families, double haploids, or RIL) is produced that can be evaluated in replicate with respect to the economic trait (Beckmann and Soller, 1986; Cowen, 1988; Ellis, 1986). In this way, environmental sources of error variation are reduced, while only one individual of each line needs to be scored with respect to the markers. Consequently, the required number of progeny groups for a given power can be less than the required number of F_2 individuals (Beckmann and Soller, 1989b).

Reducing Genetic Variance

Mapping quantitative trait loci in simple BC or F_2 populations, or their simple replicated derivatives, has one important feature that limits their usefulness: Evaluation of quantitative effects in a given marker-monitored chromosomal region takes place within the context of a variable genetic background, requiring therefore that large populations be handled for marker assessments. These drawbacks can be overcome by the use of backcross inbred lines (BIL), initially developed by Wehrhahn and Allard (1965) for the study of polygenic traits. In this design a series of lines are generated by backcrossing for several generations to the recurrent parent, before being rendered homozygous by double haploidization or selfing. The end result is a series of lines, BIL, near-isogenic to the recurrent parental line.

Wehrhahn and Allard (1965) showed that most BIL, given as few as two backcrosses, can be considered as single-gene deviates from the recurrent parent, even for quantitative traits. Upon trait evaluation, a significant fraction of the lines may show values different from that of the recurrent parent. This will allow the BIL to be separated into groups showing different performance levels. By far the largest group will represent the recurrent parent type. The deviate groups will, for the most part, represent single-gene deviates, within which the effect of each ETL (i.e., magnitude pleiotropy) can be determined precisely.

The distribution among the BIL of markers differentiating the two parents can be determined, leading to genetic characterization of the retained donor segments in each of these lines. Each BIL would be analogous to a "deletion" chromosome for mapping purposes, except that instead of being deleted, a donor chromosomal block—of varying size and position—is interpolated into a recurrent parent marker background (Beckmann and Soller, 1989a,b). Comparisons of strain distribution patterns of the discriminating markers in the different single-gene deviate BIL will generate a detailed map of the region flanking the selected locus: the more distant the marker, the less frequently it will be retained, the ETL being located between the most frequently appearing markers.

THE ROAD TO CLONING PASSES THROUGH A HIGH-RESOLUTION MAP

The cloning of agronomically important loci is still, in most instances, precarious and bewildering. The classical paradigm involves identifying a gene's product and then using this anchoring point to trace one's way back to the gene. This, however, is feasible in only a limited number of cases. In contrast, insertional mutagenesis (or transposon tagging) appears to be an attractive option for gene cloning. Geneticists are quite familiar with these techniques, which have already yielded spectacular results in many organisms (for a survey of this area see, for instance, Vodkin, 1989). Insertion of unrelated sequences inside a gene will often have deleterious effects on this gene's activity. Following recognition of the affected phenotype in a particular individual, the interrupted or uninterrupted gene can easily be retrieved. Previously, we proposed that this principle could be used as a general tool for the cloning of numerous genes in plants (Soller and Beckmann, 1987). The main difficulty then was lack of an adequate experimental system. This obstacle has now been removed. Biologically active transposons have been isolated from a number of plants and successfully transferred into and activated in heterologous crops. This opens up new and exciting horizons for the cloning of specific loci in plants.

An alternative approach relies on the a priori knowledge of the chromosomal map location of the gene of interest and walking back to the gene by means of what is now commonly referred to as "reverse genetics." This implies that the exact positioning and orientation of the target gene with respect to the flanking markers is well established, and that markers in close vicinity to this gene can be identified, i.e., within walking rather than jumping distances, or, in genetic terms within a fraction of a centimorgan. Finally, adequate means need to be available to enable the recognition of the gene, once in hand. The latter goal can be taken care of by assaying the cloned gene in transgenic individuals. The former two goals require adequate marker saturation around the locus of interest and precise positioning, i.e., knowledge of the exact relative chromosomal order of the markers. Two recent proposals are summarized in the following sections.

Targeting Markers to Specific Chromosomal Regions

A molecular geneticist can, in some instances, accelerate the pace with which adequate marker coverage for specific chromosomal regions can be achieved, by appropriate exploitation of the existing breeding germplasm. This is well illustrated in the studies of Young et al. (1988) and Sarfatti et al. (1989) in which they used families of near-isogenic tomato lines (NIL)—with or without the specific locus of interest (in this case, *Tm–2a* and I2, conferring resistance to tobacco mosaic virus and to *Fusarium*, respectively)—which breeders had previously generated. Each pair of NIL differed as little as possible from the others except for the chromosomal region of interest. Such pairwise comparisons enabled these investigators to generate detailed maps around these loci.

The second approach would take advantage of the unique attributes of BIL, which to a molecular geneticist would be the equivalent of a genomic library, cloned in the recurrent parent. Indeed, as for NIL (which can be considered a particular case) a set of BIL carrying any given gene (or ETL) should enable exact alignment of the markers with respect to one another, and of the gene(s) of interest with respect to the markers. Since the genetic resolution power of BIL is greatly increased in comparison with alternative methods (Beckmann and Soller, 1989a,b), this will lead to the possibility of uncovering flanking RFLP markers in the close vicinity (a fraction of a centimorgan) of the target gene(s). These are essential features or molecular cloning of the loci governing the traits of interest via reverse genetics.

FUTURE DEVELOPMENTS

The coming decade, will see a plethora of information on marker and trait loci for a number of agricultural commodities. This knowledge will no doubt open many opportunities for the more efficient exploration and utilization of existing or even exotic germplasms. It will enable one to combat intelligently the existing trend toward a narrowing of the genetic base of the cultivated varieties. It is useful to keep in mind that even for such a crop as maize, in temperate zones of the Northern hemisphere breeders are probably sampling only about 5% of this species' germplasm (Goodman, 1985). Changing this trend will require powerful means. These are now becoming available. The main question will be what do we do next? How can one actually implement this knowledge into breeding practice?

So far, theoretical and experimental efforts in marker-based methodologies have been directed mostly at the genetic analysis of traits of interest. We now have adequate experimental designs and modes of analysis, or at least good ideas, along these lines. This is in contrast with the current lack of theory with respect to means of transforming the information generated into more effective improvement programs. (Specific applications were discussed elsewhere: Beckmann and Soller, 1986; Tanksley et al., 1989.) Implementation of marker-based methodologies in actual breeding practice is the challenge for the modern breeder, who no doubt will combine both biometrical and molecular genetic skills. Adequate designs will be required to account for a number of needs, such as those stemming from the fact that (1) breeders manipulate several traits simultaneously (Nienhuis and Helentjaris, 1989); (2) a locus having a positive effect on one trait might have negative pleiotropic effects on other traits, and thus the effects of each locus on all traits it influences need to be weighed in the breeding designs; or (3) selection in favor of some loci has to be accompanied by selection against others, at least to remove excess unwanted genetic material (linkage drag) using narrow bracketing markers.

Once all this has been done, the breeder will be able to adapt varieties to suit the local market or niches, to rapidly adjust his varieties as market demands fluctuate, and more important, to provide the adequate and rapid means to fight unexpected environmental pressures, as they originate.

The genetic and population structures of plants and animals show many similarities, and important differences too. Selfing is a common reproductive mode for many plants. Consequently, inbred lines are commonly available. This is probably the single most important reason that plants often offer much more tractable genetic systems. Hence, it is no surprise that plants may lend themselves to faster application of marker-based methodologies. It is thus likely that animal geneticists may look for inspiration in the studies and achievements made by plant scientists. We hope that the present discussion may have helped the animal scientist to achieve a clearer view of the present status of plant marker-based methodologies.

ACKNOWLEDGMENT

I extend my appreciation to my long-time collaborator, partner, and friend, Prof. Morris Soller, for a highly stimulating working relationship. His comments and help to this manuscript are greatly appreciated. So are the discussions with Yechezkel Kashi and Dr. David Gidoni. This work was supported by grants from the U.S.-Israel Binational Agricultural Research and Development Fund (BARD) and from the NCRD (Jerusalem)-GSF (Munich).

REFERENCES

Beckmann JS. Oligonucleotide polymorphisms: a new tool in genomic genetics. Bio/Technology 1988; 6:1061-1064.

Beckmann JS, Bar-Joseph M. The use of synthetic DNA probes in breeders' rights protection: a proposal to superimpose an alpha-numerical code on the DNA. Trends Biotechnol 1986; 4:230-232.

Beckmann JS, Soller M. Restriction fragment length polymorphisms in genetic improvement: methodologies, mapping and costs. Theor Appl Genet 1983; 67:35-43.

Beckmann JS, Soller M. Restriction fragment length polymorphisms in plant genetic improvement. In: Miflin BJ, ed. Oxford surveys of plant molecular and cell biology, Vol 3: Oxford: Oxford University Press, 1986: 196-250.

Beckmann JS, and Soller M. Backcross inbred lines for mapping and cloning of loci of interest. In: Development and application of molecular markers to problems in plant genetics. Helentjaris T, Burr B, eds. Cold Spring Harbor, 1989a: 117-122.

Beckmann JS, Soller M. Genomic genetics in plant breeding. Proc. XIIth EUCARPIA Congress. In: Vortr. Pflanzenzücht, Vol 16. Berlin and Hamburg: Paul Parey Science Publishers, 1989b: 91-106.

Bonierbale MW, Plaisted RL, Tanksley SD. RFLP maps based on a common set of clones reveal modes of chromosomal evolution in potato and tomato Genetics 1988; 120:1095-1103.

Botstein D, White R, Skolnick M, Davis RW. Construction of a genetic linkage map in man using restriction fragment length polymorphisms. Am J Hum Genet 1980; 32:314-331.

Burke DT, Carle GF, Olson MV. Cloning of large segments of exogenous DNA into yeast by means of artificial vectors. Science 1987; 236:806–813.

Burr B, Evola SV, Burr FA, Beckmann JS. The application of restriction fragment length polymorphism to plant breeding. In: Genetic engineering: principles and methods, Setlow JK, Hollaender A, eds. New York: Plenum Press, Vol 5. 1983: 45–59.

Burr B., Burr FA, Thompson KH, Albertsen MC, Stuber CW. Gene mapping with recombinant inbreds in maize. Genetics 1988; 118:519- 526.

Chang C, Bowman JL. Dejohn AW. Lander ES. Meyerowitz EM. Restriction fragment length polymorphisms map for *Arabidopsis thaliana*. Proc Natl Acad Sci USA 1988; 85:6856–6860.

Cowen NM. The use of replicated progenies in marker-based mapping of ETLs. Theor Appl Genet 1988; 75:857–862.

Dallas JF. Detection of DNA "fingerprints" of cultivated rice by hybridization with a human minisatellite probe. Proc Natl Acad Sci USA 1988; 85:6831–6835.

Edwards MD, Stuber CW, Wendel JF. Molecular-marker-facilitated investigations of quantitative-trait loci in maize. I. Numbers, genomic distribution and types of gene action. Genetics 1987; 116:113–125.

Ellis THN. Restriction fragment length polymorphism markers in relation to quantitative characters. Theor Appl Genet 1986; 72:1–2.

Evola SV, Burr FA, Burr B. The suitability of restriction fragment length polymorphisms as genetic markers in maize. Theor Appl Genet 1986; 71:765–771.

Gebhardt C, Ritter E, Debener T, Schachtsnabel U, Walkemeier B, Uhrig H, Salamini F. RFLP analysis and linkage mapping in *Solanum tuberosum*. Theor Appl Genet 1989; 78:65–75.

Goodman MM. Exotic maize germplasms: status, prospects, and remedies. Iowa 1985; 59:497–527.

Grant D, Blair D, Behrendsen W, Meier R, Beavis B, Bowen S, Tenborg R, Martich J, Fincher R, Smith S, Smith H, Keashall J. Identification of quantitative-trait loci for plant height and ear height using RFLPs. Maize Genet Coop Newslett 1988; 62:73–75.

Grodzicker T, Williams J, Sharp P. Sambrook, J. Physical mapping of temperature-sensitive mutations of adenoviruses. Cold Spring Harbor Symp Quant Biol 1974; 39:439–446.

Hamada H, Petrino MG Kakunaga T. A novel repeated element with Z-forming potential is widely found in evolutionary diverse eukaryotic genomes. Proc Natl Acad Sci USA 1982; 6465–6469.

Haughe BM, Giraudat J, Nam H.-G. Goodman HM. Progress toward a physical map of the *Arabidopsis thaliana* genome. In: Development and application of molecular markers to problems in plant genetics. Helentjaris T. Burr, B. eds. New York: Cold Spring Harbor, 1989:149-151.

Havey MJ, Muehlbauer FJ. Linkages between restriction fragment length, isozyme, and morphological markers in lentil. Theor Appl Genet 1989; 77:395-401.

Helentjaris T, Burr B. Development and application of molecular markers to problems in plant genetics. New York: Cold Spring Harbor, 1989: 165 pp.

Helentjaris T, Slocum M, Wright S, Schaeffer A, Nienhuis J. Construction of genetic linkage maps in maize and tomato using restriction fragment length polymorphisms. Theor Appl Genet 1986: 72:761-769.

Helentjaris T, Weber D, Wright S. Identification of the genomic locations of duplicate nucleotide sequences in maize by analysis of restriction fragment length polymorphisms. Genetics 1988; 118:353- 363.

Jeffreys AJ. Highly variable minisatellites and DNA fingerprints. Biochem Soc Trans 1987; 15:309-317.

Jeffreys AJ, Wilson V, Thein SL. Hypervariable "minisatellite" regions in human DNA. Nature 1985; 314:67-73.

Kahler AL, Wehrhahn CF. Associations between quantitative traits and enzyme loci in the F_2 population of a maize hybrid. Theor Appl Genet 1986; 72:15-26.

Kashi Y. Soller M, Hallerman E, Beckmann JS. Restriction fragment length polymorphisms in dairy cattle genetic improvement. In: Proc. of the Third World Congress on Genetics Applied to Livestock Production, Nebraska, 1986, Vol 12. 1986: 57-63.

Kashi Y, Iraqi F, Tikoschinksy Y, Ruzinski B, Nave A, Beckmann JS, Friedman A, Soller M, Gruenbaum Y. $(TG)_n$ uncovers a sex-specific hybridization pattern in cattle. Genomics 1990a; 6:31-36.

Kashi Y, Tikoschinksy Y, Genislav E, Iraqi F, Nave A, Beckmann JS, Gruenbaum Y, Soller M. Large restriction fragments containing poly-TG are highly polymorphic in a variety of vertebrates. Nucl Acids Res 1990b; 18:1129-1132.

Lander ES, Botstein D. Mapping mendelian factors underlying quantitative traits using RFLP linkage maps. Genetics 1989; 121:185-199.

Landry BS, Kesseli RV, Farrara B, Michelmore RW. A genetic map of lettuce (*Lactuca sativa*) with restriction fragment length polymorphism, isozyme, disease resistance and morphological markers. Genetics 1987; 116:331-337.

Lebowitz RJ, Soller M, Beckmann JS, Trait-based analyses for the detection of linkage between marker loci and quantitative trait loci in crosses between inbred lines. Theor Appl Genet 1986; 72:556-562.

Lee M, Godshalk EB, Lamkey KR, Woodman WW, Association of restriction fragment length polymorphisms among maize inbreds with agronomic performance of their crosses. Crop Sci 1989; 29:1067-1071.

Litt M, Luty JA. A hypervariable microsatellite revealed by in vitro amplification of a dinucleotide repeat within the cardiac muscle actin gene. Am J Hum Genet 1989; 44:397-401.

Martin B, Nienhuis J, King G, Schaefer A. Restriction fragment length polymorphisms associated with water use efficiency in tomato. Science 1989; 243:1725-1728.

McCouch SR, Kochert G, Yu ZH, Wang ZY, Kush GS, Coffman WR, Tanksley SD. Molecular mapping of rice chromosomes. Theor Appl Genet 1988; 76:815-829.

Meyerowitz EM. *Arabidopsis*, a useful weed. Cell 1989; 56:263-269.

Nadeau JH. Maps of linkage and synteny homologies between mouse and man. Trends Genet 1989; 5:82-86.

Nakamura Y, Lepper M, O'Connel P, Wolff R, Holm T, Culver M, Martin C, Fujimoto E, Hoff M, Kumlin E, White R. Variable number of tandem repeat (VNTR) markers for human gene mapping. Science 1987; 235:1616-1622.

Nienhuis J, Helentjaris T. Simultaneous selection for multiple polygenic traits through RFLP analysis. In: Development and application of molecular markers to problems in plant genetics. Helentjaris T, Burr B. New York: Cold Spring Harbor, 1989:107-112.

Nienhuis J, Helentjaris T, Slocum M, Ruggero B, Schaefer A. Restriction fragment length polymorphism analysis of loci associated with insect resistance in tomato. Crop Sci 1987; 27:797- 803.

Nybom H, Rogstadt SH, Schaal BA. Genetic variation detected by use of M13 "DNA fingerprint" probe in *Malus, Prunus and Rubus* (Rosaceae). Theor Appl Genet 1990; 79:153-156.

Osborn TC, Alexander DC, Fobes JF. Identification of restriction fragment length polymorphisms linked to genes controlling soluble solids content in tomato fruits. Theor Appl Genet 1987; 73:350-356.

Paterson AH, Lander ES, Hewitt JD, Peterson S, Lincoln SE, Tanksley SD. Resolution of quantitative traits into mendelian factors by using a complete linkage map of restriction fragment length polymorphisms. Nature 1988; 335:721-726.

Poustka A, Lehrach H. Jumping libraries and linking libraries: the next generation of molecular tools in mammalian genetics. Trends Genet 1986; 2:174–179.

Rick C. Controlled introgression of chromosomes of *Solanum pennellii* into *Lycopersicon esculentum*: segregation and recombination. Genetics 1969; 62:753–768.

Rogstadt SH, Patton JC, Schaal BA. M13 repeat probe detects DNA minisatellite-like sequences in gymnosperms and angiosperms. Proc Natl Acad Sci USA 1988; 85:9176–9178.

Rom M. Development of restriction fragment length polymorphism in plants as a tool for strain identification and breeding (Hebrew, with summary in English). M.Sc. thesis Hebrew University of Jerusalem, Israel, May 1982.

Royle NJ. Clarkson RE, Wong Z, Jeffreys AJ. Clustering of hypervariable minisatellites in the proterminal regions of human autosomes. Genomics 1988; 3:352–360.

Ryskov AP, Jincharadze AG, Prosnyak MI, Ivanov PL, Limborska SS. M13 phage DNA as a universal marker for DNA fingerprinting of animals, plants and microorganisms. FEBS Lett 1988; 233:388–392.

Saiki RK, Gelfand DH, Stoffel S, Scharf SJ, Higuchi R, Horn GT, Mullis KB, Erlich HA. Primer-directed enzymatic amplification of DNA with a thermostable DNA polymerase. Science 1988; 239:487– 491.

Sarfatti M, Katan J, Fluhr R, Zamir D, An RFLP marker in tomato linked to the *Fusarium oxysporum* resistance gene I2. Theor Appl Genet 1989; 78:755–760.

Sharp PJ, Chao S, Deasi S, Gale MD. The isolation, characterization and application in the *Triticeae* of a set of wheat RFLP probes identifying each homeologous chromosome arm. Theor Appl Genet. 1989; 78:342–348.

Slocum MK. Analyzing the genomic structure of *Brassica* species and subspecies using RFLP analysis. In: Development and application of molecular markers to problems in plant genetics. Helentjaris T, Burr B, eds. New York: Cold Spring Harbor, 1989:73–80.

Smith CL, Cantor CR, Approaches to physical mapping of the human genome. Cold Spring Harbor Symp Quant Biol 1987; 51:115–122.

Soller M, Beckmann JS. Restriction fragment length polymorphisms and genetic improvement. In: Proceedings of the Second World Congress on Genetics Applied to Livestock Production, Madrid, 1982, Vol 6. 1982: 396–404.

Soller M, Beckmann JS. Genetic polymorphism in varietal identification and genetic improvement. Theor Appl Genet 1983; 67:25–33.

Soller M, Beckmann JS. Cloning of quantitative trait loci by means of insertional mutagenesis. Theor Appl Genet 1987; 74:369-378.

Soller M, Beckmann JS. Genomic genetics and the utilization for breeding purposes of genetic variation between populations. In: Proceedings of the Second International Conference on Quantitative Genetics. Weir BS, Eisen EJ, Goodman MM, Namkoong G, eds Sunderland Sinauer, MA: 1988: 161-188.

Soller M, Genizi A, Brody T. On the power of experimental designs for the detection of linkage between marker loci and quantitative loci in crosses between inbred lines. Theor Appl Genet 1976; 47:35- 39.

Song KM, Osborn TC, Williams PH. *Brassica* taxonomy based on nuclear restriction fragment length polymorphisms (RFLPs) 1. Genome evolution of the diploid and amphidiploid species. Theor Appl Genet 1988; 75:784-794.

Stuber CW, Edwards MD, Wendel JF. Molecular marker-facilitated investigations of quantitative trait loci in maize. II. Factors influencing yield and its component traits. Crop Sci 1987; 639-648.

Tankley SD, Hewitt J. Use of molecular markers in breeding for soluble solids content in tomato—a re-examination. Theor Appl Genet 1988; 75:811-823.

Tanksley SD, Medina-Filho H, Rick CM. Use of naturally-occurring enzyme variation to detect and map genes controlling quantitative traits in an interspecific backcross of tomato. Heredity 1982; 11-25.

Tanksley SD, Bernatzky R, Lapitan NL, Price JP. Conservation of gene repertoire but not gene order in pepper and tomato. Proc Natl Acad Sci USA 1988a; 85:6419-6423.

Tanksley SD, Miller J, Paterson A, Bernatzky R. Molecular mapping of plant chromosomes. In: Gustafson JF. Appels R, eds. Chromosome structure and function. New York: Plenum Press, 1988b:157-173.

Tanksley SD, Young ND, Paterson AH, Bonierbale MW. RFLP mapping in plant breeding: new tools for an old science. Bio/Technology 1989; 7:257-266.

Vallejos CE, Tanksley SD. Segregation of isozyme markers and cold tolerance in an interspecific backcross of tomato. Theor Appl Genet 1983; 66:241-247.

Vassart T, Georges M, Monsieur R, Brocas H, Lequarre AS, Christophe D. A sequence in M13 phage detects hypervariable minisatellites in human and animal DNA. Science 1987; 235:683-684.

Vodkin LO. Transposable element influence on plant gene expression and variation. In: Marcus A, Stumpf PK, Conn EE, eds. The biochemistry of plants: a comprehensive treatise. San Diego: Academic Press, Vol 15. 1989:83-132.

Weber D, Helentjaris T. Mapping RFLP loci in maize using B-A translocations. Genetics 1989; 121:583-590.

Weber JL, May PE. Abundant class of human DNA polymorphisms which can be typed using the polymerase chain reaction. Am J Hum Genet 1989; 44:388-395.

Wehrhahn C, Allard R. The detection and measurement of the effects of individual genes involved in the inheritance of a quantitative character in wheat. Genetics 1965; 51:109-119.

Weller JI, Soller M, Brody T. Linkage analysis of quantitative traits in an interspecific cross of tomato *Lycopersicon esculentum* x *Lycopersicon pimpinellifolium*) by means of genetic markers. Genetics 1988; 118:329-339.

Womack JE, Moll YD. Gene map of the cow: conservation of linkage with mouse and man. J Hered 1986; 77:2-7.

Young ND, Zamir D, Ganal MW, Tanksley SD. Use of isogenic lines and simultaneous probing to identify DNA markers tightly linked to the *Tm-2a* gene in tomato genetics 1988; 120:579-585.

11

Restriction Fragment Length Polymorphism Applications in Human Chromosome Mapping

Paul C. Watkins

Life Technologies, Inc.
Gaithersburg, Maryland

INTRODUCTION

As a first step toward obtaining a complete description of the human genome, detailed physical and genetic linkage maps of all the human chromosomes must be established. Since the idea was first proposed in 1980 that a genetic linkage map of the human genome could be made using DNA probes that detect restriction fragment length polymorphisms (RFLPs) (1), an extraordinary amount of progress has been made in developing DNA markers and using them to map human chromosomes.

Today, preliminary genetic linkage maps exist for all the human chromosomes. A few chromosomes and specific chromosome regions, especially those near genetic disease loci, have been mapped with RFLPs to a high density. With detailed genetic linkage maps, researchers will be able to identify the location of genetic disease genes far more efficiently than they have been able to in the past. In earlier linkage studies, the chromosome location of the polymorphic DNA probe, and therefore the linked genetic disease locus, was determined only after linkage between a DNA marker and the genetic disease was first established. This strategy relied on testing dozens and potentially hundreds of RFLPs in a random fashion until linkage was detected. Although several important genetic disease genes such as those causing Huntington's disease on chromosome 4 (2) and cystic fibrosis on chromosome 7 (3–5), were successfully localized to specific chromosomes using this strategy, it is very inefficient and depends a great deal on the lucky choice of the correct probe for linkage testing. Once sets of DNA markers and high-resolution maps of all the chromosomes become available, a systematic approach to searching for genes causing genetic disease can be employed. Entire chromosomes can be tested for linkage and excluded, or if linkage is detected, the map position of the genetic disease gene would be automatically determined.

The efficient construction of human genetic linkage maps requires highly polymorphic DNA markers, DNA from a large number of families, and computerized methods to perform the linkage analysis. Large numbers of polymorphic DNA markers are needed in order to have a high probability of finding a set of markers that span the length of a chromosome with sufficient density so that gaps in the map are not too large. Since linkage maps are based on meiotic recombination, large numbers of individuals must be analyzed in order to have enough crossover events so that markers can be ordered with a high degree of confidence. Finally, efficient computer programs are required so that the enormous amount of data generated during the linkage study can be organized and analyzed.

ELEMENTS OF GENETIC LINKAGE MAPS

DNA Markers

Perhaps the most significant component of the human chromosome genetic linkage map is the genetic marker that defines each point on the map. The earliest type of genetic marker used in human genetics was the protein polymorphism. This was a polymorphism of the gene product and was defined by various marker systems that included serological tests for blood group antigens and tests to detect electrophoretic variants of different proteins and enzymes. In the early 1980s the protein polymorphism was rapidly replaced by a far more useful kind of genetic marker, the RFLP. Since RFLPs are revealed by using cloned DNA probes to detect DNA sequence variation in the genome directly, two significant advantages over protein polymorphisms were obtained. First, the DNA probes that detected RFLPs could be easily localized to specific chromosomes by physical methods. Second, the development of DNA markers for specific human chromosomes could be approached in a systematic fashion. Linkage maps of all the chromosomes could be made by analyzing the inheritance pattern of the RFLPs in human pedigrees.

One of the first papers describing the construction of genetic linkage maps of chromosomes using RFLPs was published in 1985 (6). At that time, White and colleagues noted that over 200 polymorphic DNA markers had been reported. Today, over 2000 polymorphic DNA markers have been characterized and are recorded at the Howard Hughes Medical Institute (HHMI) Human Gene Mapping Library at Yale University. Over 400 of these markers are cloned cDNAs or genomic sequences associated with genes. The rest are "anonymous" DNA sequences that have been characterized as RFLP-detecting DNA markers, but it is not known which of these cloned probes contain DNA sequences from coding regions.

Although the first reported DNA polymorphism (7) was found to result from a variable insertion or deletion of tracts of DNA sequence and as a result was highly polymorphic (>10 alleles), the majority of RFLPs discovered since then are simple single-site mutations in the DNA sequence comprising the recognition site of a restriction enzyme. This type of polymorphism results in two alleles, and the degree of polymorphism, or usefulness of the RFLP as a genetic marker, is dependent on the frequency of the alleles in the population. An RFLP that has two alleles is maximally informative if both alleles are equally distributed in the population. The frequency of heterozygosity of this type of marker, i.e., the proportion of the population predicted to be heterozygotes, can never be more than 50%.

RFLPs that are close together, for example, a DNA probe that detects multiple RFLPs when different enzymes are used, can be combined and a haplotype generated for that marker. The RFLPs associated with a DNA probe that detects both an Eco R1 RFLP (allele 1 = +Eco R1 site, allele 2 = -Eco R1 site), and a Hind III RFLP (allele 1 = +Hind III site, allele 2 = -Hind III site) can be combined to generate a haplotype of four alleles (++, +-, -+, --). Combining RFLPs to form haplotypes can substantially increase the frequency of heterozygosity of a marker provided that the mutations at the restriction sites that define the haplotype are segregating completely at random in the population. For a marker with four alleles, the maximum frequency of heterozygosity is 75%.

The best DNA markers are those with multiple alleles, especially when the alleles are equally distributed in the population. The likelihood that an individual in a linkage study will be heterozygous for such a marker is very high. Recent attention has focused on a type of DNA marker that results from a variable number of tandem repeats of short stretches of DNA sequence. This kind of marker (termed a VNTR marker), unlike a single restriction site polymorphism, will exhibit a large number of different restriction fragment lengths as a result of the highly variable number of the tandem repeat units found at a genetic locus. Although the VNTR marker is similar to the multilocus minisatellite probes used for DNA fingerprinting, the VNTR probe represents polymorphism at a single genetic locus and is therefore much more easily used in a linkage study. Nakamura and colleagues have successfully isolated and characterized a number of new VNTR markers by screening human genomic libraries with synthetic oligonucleotide DNA probes based on the reported DNA repeat sequence of several known VNTR markers (8). Because of their high degree of polymorphism, the VNTR markers have had a significant impact on the construction of linkage maps of human chromosomes.

The development of DNA markers for linkage mapping has proceeded at a rapid pace. A conventional approach to screening DNA probes for the ability to detect RFLP has evolved. Virtually any single-copy DNA sequence can be hybridized to genomic DNA digested with 15-20 restriction enzymes from four to five unrelated individuals with a reasonable chance of detecting an RFLP. In the author's laboratory, approximately one of every five clones isolated from a library of human chromosome 21 detected an RFLP (9). Some groups have established conditions using clones prehybridized with unlabeled human DNA to eliminate interference from repetitive sequences, thereby allowing entire lambda or cosmid clones to be screened for RFLP-detecting ability without having to first identify and subclone single-copy regions (10).

A significant advance in the development of chromosome-specific DNA markers, and one that will accelerate the pace of chromosome linkage mapping, is the construction of DNA libraries from flow-sorted chromosomes (11,12). An effort coordinated under the auspices of the National Laboratory Gene Library Project begun in 1983 has resulted in the creation of DNA libraries from human chromosomes purified by fluorescence-activated flow sorting (13). High-density linkage maps of human chromosomes, such as the one reported for chromosome 7 (14), have been made from DNA markers obtained after screening one of these libraries. DNA libraries of all the human chromosomes are now available through the American Type Culture Collection. Although the existing libraries contain relatively small inserts in a lambda phage vector, newer libraries made with large inserts in cosmid vectors are currently under construction.

Families

The efficiency of map construction and the precision of the map are highly dependent on the pedigrees analyzed in the linkage study. Just as highly polymorphic markers increase mapping efficiency by increasing the proportion of individuals who are heterozygous at the marker locus, family structure can greatly influence the amount and quality of information gained from a linkage study. Families that are ideally suited for mapping have large numbers of offspring and include both sets of grandparents. It is easier and more likely that the phase, the combination of alleles on the chromosomes of the parents, can be established given this kind of family. It is also more efficient to sample fewer parental and grandparental DNAs if the family sizes are large.

Many of the earlier linkage studies were exclusively directed toward finding DNA markers linked to genetic disease loci. Families used for these studies were, of course, selected on the basis of having the inherited disease. With a few exceptions, such as the Venezuelan reference pedigrees used to study Huntington's disease (2), most of the families selected for genetic disease studied are not well suited for making linkage maps of chromosomes.

In 1984 an international collaboration, The Centre d'Etude du Polymorphisme Humain (CEPH), was established to provide investigators with a valuable mapping resource: a centralized bank of DNA from lymphoblastoid cell lines established from a number of large reference pedigrees. The CEPH maintains cell lines and coordinates the distribution of DNA from 40 three-generation human families. The families include grandparents and have an average of eight children.

As genetic markers are typed against the CEPH families, the data are compiled into a data base. This enables laboratories not only to generate and test their markers independently, but to combine data on the same sets of families in order to establish the chromosome linkage map. Since crossovers between all the markers are defined on the same sets of chromosomes, linkage maps can be developed with new markers oriented with respect to previously established crossover points. Most of the published genetic linkage maps of human chromosomes have been established using either a subset or all of the 40 families currently maintained by CEPH. Twenty additional large families have been identified and will soon be included in this resource.

Analytical Methods

Given a set of polymorphic markers and suitable families for testing, a genetic linkage map can be made by analyzing the inheritance pattern of the markers through all the family members. Since many markers and individuals are employed in the analysis, computer programs have been created to handle the computation-intensive requirements of linkage map construction. Maps are based on the principle of meiotic recombination; marker loci that are close together on a chromosome are less frequently separated by recombination, or chromosome crossover events, than are marker loci father apart on a chromosome.

A complete description of the methods used in the analysis of human genetic linkage is given by Ott (15). There are several objectives when developing a chromosome genetic linkage map starting with a set of chromosome-specific DNA markers: (1) Markers are first tested in all pairwise combinations to find groups of linked markers. (2) Marker order is established within the linkage groups. Marker order and group order are determined by multilocus analysis, and the distance between markers is reported as recombination fraction or as map units by using a mapping function to convert recombination fraction to genetic map distance. (3) The statistical support for the most likely marker order is determined and compared to alternative orders. (4) The orientation of the genetic linkage map with respect to the physical chromosome map is established by high-resolution physical mapping of some of the markers. (5) During the analysis, male or female meioses can be evaluated independently of the other sex, and separate genetic maps based on male or female recombination frequency can be established.

Several linkage mapping strategies have been developed and form the basis for computer programs. One of the earliest linkage programs, LIPED (16), although widely used especially for genetic disease analysis, is limited to analyzing pairs of marker loci. Newer linkage programs such as LINKAGE (17), CRI-MAP and MAPMAKER (18), and GMS (19) have been developed for construction of multilocus chromosome linkage maps from complex data sets. Modifications to these programs make them especially well suited for analyzing data generated on the CEPH pedigrees.

The human genome is approximately 3300 centimorgans (cM) in genetic length (1 cM = 1% recombination). Recombination estimates for intervals between markers on a map are usually converted to map units and reported as centimorgans. A mapping function must be used to convert recombination fraction to centimorgans. The choice of a mapping function depends on an assumption about the presence of interference, the influence of one chromosome crossover event (chiasma) on a subsequent crossover. Generally, maps made using LINKAGE convert recombination fraction to centimorgans by using Haldane's mapping function which assumes an interference level of zero. Maps made using CRI-MAP and MAPMAKER use a different mapping function (Kosambi's) for the conversion of recombination fraction to map units under the assumption of a moderate level of interference. Genetic linkage maps that are derived from the same data set but reported in Kosambi centimorgans will be slightly smaller than those reported in Haldane centimorgans [see Ott (15) for a detailed explanation of mapping functions].

Once order and distances are established for a set of markers, the statistical support for the preferred order over possible alternative orders is calculated. Most investigators test the preferred marker order by calculating the odds against inverting adjacent marker loci. The odds are reported for each marker interval. Support for a particular marker order is influenced by a number of factors, including the degree of polymorphism of the markers, the number of meioses contributing to the analysis, and the distance of the markers from one another.

With advances in physical mapping methods, genetic markers can be precisely localized to specific physical regions of the chromosome, thereby providing anchor points for orientation of the genetic linkage map to the physical chromosome map. DNA probes can be localized to specific chromosome regions by direct hybridization to metaphase chromosomes (in situ hybridization). Alternatively, a Southern blot can be performed using a collection of somatic cell hybrids, each hybrid containing a specific subset of human chromosomes. If a collection of hybrid cell lines contains different translocations of the chromosome of interest, probes can be ordered relative to the translocation breakpoints. Structural features of the human karyotype revealed as Giemsa (G), quinacrine (Q), reverse (R), and centromeric (C) bands [see Therman (20) for a description of human chromosomes] are used as the cytogenetic landmarks to localize DNA probes. Detailed cytogenetic maps of the chromosomes can be developed when banding techniques are combined with methods to obtain elongated chromosomes that display up to 1200 bands in a karyotype (21,22).

STATUS OF HUMAN CHROMOSOME GENETIC LINKAGE MAPS

Preliminary or primary genetic linkage maps have been established for all the human autosomes and the X chromosome. In many cases, several investigators have independently published linkage maps for the same chromosome (chromosomes 1, 4, 6, 7, 9, 10, 12, 13, 15, 17, 18, 21, 22, and X). Tables 1 and 2 summarize the data on the markers used to make the maps, and Table 3 compares the genetic linkage maps published to date. For the maps reported by Donis-Keller et al. (23), only markers comprising a continuous linkage group on a chromosome were included. The maps reported by this group for chromosome 14 (two separate linkage groups of 25 cM and 17 cM) and chromosome 19 (a 13-cM linkage group of five markers) were not included. The chromosome 7 map reported by Barker et al. (14) is a more detailed analysis of the data included in the original chromosome 7 map reported by Donis-Keller et al. (23) and was therefore included instead of the original. All the maps

Table 1 Markers Assigned to Genetic Maps of Human Chromosomes

Chromosome	Author (ref.)	No. of DNA Markers	No. of RFLPs	Other Markers[a]	Average Heterozygosity[b]	Cloned Genes	No. of VNTR loci[c]	Total loci[d]	No. of discrete loci[e]	Shared loci[f]	Total independent loci[g]	Genetic Disease genes (ref.)
1	Dracopoli (42)	24	29	RH, FY, PGM1	41%	PND, ALPL, MYCL FUCA1, SRC2, GLUT TSHB, NGFB	0	27	25			
1	O'Connell (31)	24	26	F13B, PGM1, RH, FY	52%	NGFB, NRAS, HTF SPTA1, PUM, REN	10	28	26			
1	Donis-Keller (23)	27	30	RH, FY, PGM1	46%	AT3, REN	NR	30	27			
										D1S2, D1S57, NGFB PND, RH, FY, PGM1	74	heredtiary cutaneous malignant melanoma (29)
2	Donis-Keller (23)	27	30		44%	CRYGP1	NR	27	22			
3	Doniş-Keller (23)	17	19		51%		NR	17	15			
4	Donis-Keller (23)	17	19	GC, MNS	44%	ADH3	NR	19	17			
4	Murray (24)	10	13	GC, MNS		ALB, MT2P1, INP10 IL2, FGB, FGG, EGF	0	12	10	GC, MNS	29	Huntington's disease (2)
5	Donis-Keller (23)	28	31		48%	DHFR, HEXB, HPRTP2	NR	28	21			famial polyposis (43)
6	Donis-Keller (23)	17	19	GLO1	43%	HLA-B, HLA-DQA HLA-DRA1	0	17	14			
6	Leach (44)	6	7	GLO1	NR	HLA-B, HLA-DQA HLA-DRA1	0	7	4	D6S5, D6S8, D6S10, HLA-B, HLA-DQA, HLA-DRA1, GLO1	17	
7	Barker (14)	63	70		42%	MET, COL1A2, TCRB	0	NR	63	35		
7	Lathrop (25)	6	11	CF	56%	MET, COL1A2, TCRB	0	7	7	D7S8, COL1A2, MET,	65	cystic fibrosis (CF) (3–5)
8	Donis-Keller (23)	14	16		50%	PLAT	NR	14	11			
9	Donis-Keller (23)	7	8	ABO, AK1, ORM	41%	ABL	NR	10	9			
9	Lathrop (32)	15	15	ABO, AK1, ORM GALT	41%	ABL, ASSP3	7	19	17	D9S1, ABL, AK1 ABO, ORM	24	
10	Lathrop (19)	11	13		43%	RBP3	1	11	10			
10	Farrer (28)	5	5	MEN2A	56%	RBP3	0	6	6			
10	Donis-Keller (23)	8	9		44%		NR	8	7	RBP3, D10S4	22	multiple endocrine neoplasia type 2A (MEN2A) (45-46)

Table 1 (Continued)

Chromo-some	Author (ref.)	No. of DNA Markers	No. of RFLPs	Other Markers[a]	Average Hetero-zygosity[b]	Cloned Genes	No. of VNTR loci[c]	Total loci[d]	No. of discrete loci[e]	Shared loci[f]	Total inde-pendent loci[g]	Genetic Disease genes (ref.)
11	Donis-Keller (23)	21	27			HBBC, HRAS1 INS, SEA	NR	21	17			bipolar affective disorder (47) multiple endocrine neoplasia type 1 (MEN1) (48)
12	O'Connell (33)	16	20		44%	KRAS2, PRB1, PRB2 PRB3, PRB4, COL2A1 F8VWF, PAH, ELA1	0	16	14			
12	Donis-Keller (23)	10	11		46%	PRB1	NR	10	9	PRB1	25	
13	Donis-Keller (23)	10		ESD			NR	10	7			
13	Bowcock (26)	8	12	ESD, WND	54%	ESD, RB1	0	9	10			
13	Haines (27)	6	9		61%	ESD, RB1	0	6	6			
13	Leppert (49)	8	14	ESD	47%		0	8	5	D13S1, D13S10, RB1 D13S4, D13S5, D13S6 D13S7, ESD, D13S22 D13S2	14	retinoblastoma (RB1) (50) Wilson's disease (WND) (51)
14	Nakamura (34)	10	10	PI	60%	IGHJ	6	11	11			
15	Nakamura (35)	16	18		42%		1	17	14			
15	Donis-Keller (23)	9	10		47%		NR	9	9	D15S1, D15S2	24	adult polycystic kidney disease (52)
16	Donis-Keller (23)	42	NR		NR	HBA1, HP	NR	42	31			
17	Nakamura (36)	22	25		58%	MYH2, TK1	8	21	20			
17	Donis-Keller (23)	6	6		41%	MYH2, NGFR	NR	6	5	D17Z1, MYH2	25	von Recklinghausen neurofibromatosis (NF1) (53-54)
18	O'Connell (55)	11	11	JK	49%		2	12	12			
18	Donis-Keller (23)	8	9		54%		NR	8	8	D18S3, D18S1	18	myotonic dystrophy (56)
19	Nakamura (37)	10	11	LE, SE	50%	LDLR, APOC2, INSR	3	12	11			
20	Donis-Keller (23)	7	7		56%		NR	7	6			
21	Tanzi (38)	14	20		42%	SOD1, APP	0	14	10			
21	Warren (39)	17	19		49%	APP, SOD, ETS2 PFKL, BCEI, COL6A1	3	17	16			

21	Donis-Keller (23)	5	5		43%	SOD1, BCEI	1	5		D21S1, D21S11, D21S13 D21S8, D21S111, D21S17 D21S3, D21S17, APP D21S15, SOD1	22	familial Alzheimer's disease (57)
22	Rouleau (58)	13		P1		IGLV, BCR, PDGFB	0	14	13			
22	Julier (30)	5	8	P1	46%	IGLV, IGLC, PDGFB, MB	0	6	5			
22	Donis-Keller (23)	5	5		38%	PDGFB	0	5	5	IGLV, D22S10, PDGFB	22	bilateral acoustic neurofibromatosis (NF2) (59)
X	Drayna (60)	20	23	Xg	47%	F9, F8C, HPRT	0	21	14			
X	Donis-Keller (23)	16	17		48%	F9, F8C, HPRT	0	16	11			
X	Arveiler (61)	9	9		40%		0	9	8	HPRT, DXS143, DXS43 DXS41, DXS51, DXS15 DXS52, F9, F8C, DXS9 DXYS1, DXS3, DXS42	29	Duchenne muscular dystrophy (62)

NR = not reported.

[a]Symbols for protein polymorphisms are shown, descriptions are found in Table 2.

[b]The average frequency of heterozygosity for markers on a linkage map was calculated from reported observed heterozygosity or from reported allele frequencies assuming Hardy Weinberg equilibrium and that allele association was random for haplotyped markers.

[c]VNTR loci are loci defined by VNTR type markers.

[d]Total loci is the sum of DNA markers and other markers on the genetic map.

[e]Only distinct points on the linkage map were counted. If a pair or group of markers showed no recombination, they were counted as a single discrete locus.

[f]Listed are markers that have been included on at least two of the published linkage maps.

[g]Listed are the number of unique markers that have been assigned to all the published linkage maps of a given chromosome.

Table 2 Genes That Detect RFLPs and Protein Polymorphisms Assigned to Genetic Linkage Maps of Human Chromosomes

Chromo-some	Cloned genes		Protein polymorphisms	
	Symbol	Gene	Symbol	Protein
1	ALPL	alkaline phosphatase	F13B	coagulation factor XIII, beta polypeptide
1	AT3	antithrombin III		
1	FUCA1	alpha-L-1 fucosidase	FY	Duffy blood group
1	GLUT	glucose transport	PGM1	phosphoglucomutase 1
1	HTF	human tissue factor	RH	Rhesus blood group
1	MYCL	avian myelocytomatosis viral (v-myc) oncogene homolog		
1	NGFB	nerve growth factor, beta polypeptide		
1	NRAS	neuroblastoma RAS viral (v-ras) oncogene homolog		
1	PND	pronatriodilatin		
1	PUM	peanut lectin binding urinary mucin		
1	REN	renin		
1	SPTA1	alpha spectrin		
1	SRC2	Rous sarcoma virus, cellular homolog		
1	TSHB	thyroid stimulating hormone, beta polypeptide		
2	CRYGP1	crystallin gamma polypeptide pseudogene 1		
4	ADH3	alcohol dehydrogenase (class I), gamma polypeptide	GC	group-specific component (vitamin D binding protein)
4	ALB	albumin	MNS	MNS blood group
4	EGF	epidermal growth factor		
4	FGB	fibrinogen, B beta polypeptide		
4	FGG	fibrinogen, gamma polypeptide		
4	INP10	protein 10 from gamma interferon-induced cell line		
4	IL2	interleukin 2		
4	MT2P1	metallothionein 2, pseudogene 1 (processed)		
5	DJFR	dihydrofolate reductase		
5	HEXB	hexosaminidase B, beta polypeptide		
5	HPRTP2	hypoxanthine phosphoribosyltransferase pseudogene 2		
6	HLA-B	major histocompatibility complex, class I		
6	HLA-DQA	major histocompatibility complex, class II, alpha		
6	HLA-DRA1	major histocompatibility complex, class II, alpha		
7	COL1A2	collagen, type I, alpha 2		
7	MET	met proto-oncogene		
7	TCRB	T cell receptor, beta polypeptide		
8	PLAT	tissue plasminogen activator		
9	ABL	Abelson murine leukemia viral (v-abl) oncogene homolog	ABO	ABO blood group
9	ASSP3	argininosuccinate synthetase pseudogene 3	AK1	adenylate kinase 1
9			GALT	galactose-1-phosphate uridyltransferase
9			ORM1	orosomucoid 1
10	RBP3	interstitial retinol-binding protein 3		
11	HBC	beta hemoglobin cluster		
11	HRAS1	Harvey rat sarcoma 1 viral (v-Ha-ras1) oncogene homolog		
11	INS	insulin		
11	SEA	S13 avian erythroblastosis viral oncogene homolog		
12	COL2A1	collagen type II, alpha 1		
12	KRAS2	Kirsten rat sarcoma 2 viral (v-Ki-ras2) oncogene homolog		
12	PAH	phenylalanine hydroxylase		
12	PRB1	proline-rich protein BstNI subfamily 1		
12	PRB2	proline-rich protein BstNI subfamily 2		

Table 2 (Continued)

Chromosome	Cloned genes: Symbol	Cloned genes: Gene	Protein polymorphisms: Symbol	Protein polymorphisms: Protein
12	PRB3	proline-rich protein BstNI subfamily 3		
12	PRB4	proline-rich protein BstNI subfamily 4		
12	F8VWF	coagulation factor VIII VWF (von Willebrand factor)		
13	ESD	esterase D	ESD	esterase D
13	RB1	retinoblastoma 1		
14	IGHJ	immunoglobulin in heavy polypeptide, joining region	PI	alpha-1-antitrypsin
16	HBA1	alpha 1 hemoglobin		
16	HP	haptoglobin		
17	NGFR	nerve growth factor receptor		
17	MYH2	myosin, heavy polypeptide 2 skeletal muscle, adult		
17	TK1	thymidine kinase		
18			JK	Kidd blood group
19	APOC2	apolipoprotein C-II	LE	Lewis blood group
19	INSR	insulin receptor	SE	ABH secretion
19	LDLR	low density lipoprotein receptor		
21	APP	amyloid beta (A4) precursor protein		
21	BCEI	estrogen-inducible sequence, expressed in breast cancer		
21	COL6A1	collagen type VI, alpha 1		
21	ETS2	avian erythroblastosis virus E26 (v-ets) oncogene homolog 2		
21	PKFL	phosphofructokinase, liver type		
21	SOD1	superoxide dismutase		
22	BCR	breakpoint cluster region	P1	P blood group
22	IGLC	immunoglobulin lambda polypeptide, constant region		
22	IGLV	immunoglobulin lambda polypeptide, variable region		
22	MB	myoglobin		
22	PDGFB	platelet derived growth factor, beta polypeptide, (simian sarcoma viral (v-sis) oncogene homolog)		
X	F8C	coagulation factor VIIIc, procoagulant component (hemophilia A)	XG	Xg blood group
X	F9	coagulation factor IX, plasma thromboplastic component (hemophilia B)		
X	HPRT	hypoxanthine phosphoribosyltransferase		

Table 3 Comparison of Genetic Linkage Maps of Human Chromosomes

Chromosome	Author (ref.)	Map length		Average distance between loci	Largest gap	No. of gaps > 25 cm	Pedigrees (no. of families)[a]	Programs[b]	Support for locus order[c]	
		Male	Female						> 100:1	> 1000:1
1*	Dracopoli (42)	102	230	10	34	1	CEPH (40)	MAPMAKER	18/25	16/25
1	O'Connell (31)	320	608	24	115	8	CEPH (40) Non-CEPH (19)	GMS LINKAGE	23/25	22/25
1*	Donis-Keller (23)	199	371	16	50	6	CEPH (23) MAPMAKER	CRI-MAP	21/26	NI
2*	Donis-Keller (23)	140	236	12	17	0	CEPH (23)	CRI-MAP MAPMAKER	16/21	NI
3*	Donis-Keller (23)	151	235	17	28	4	CEPH (23)	CRI-MAP MAPMAKER	10/14	NI
4*	Donis-Keller (23)	108	292	19	50	5	CEPH (23)	CRI-MAP MAPMAKER	10/16	NI
4	Murray (24)	89	185	23	50	3	CEPH (40)	LINKAGE (2 point)	NI	NI
5*	Donis-Keller (23)	139	252	14	44	2	CEPH (23)	CRI-MAP MAPMAKER	16/20	NI
6*	Donis-Keller (23)	122	185	17	50	3	CEPH (23)	CRI-MAP MAPMAKER	7/12	NI
6	Leach (44)	18	24	8	9	0	Non-CEPH (25)	LINKAGE	NI	NI
7	Lathrop (25)	34	106	18	66	1	CEPH (40) Non–CEPH (53)	LINKAGE	2/3	2/3
7*	Barker (14)	170	250	9	40	2	CEPH (23)	CRI-MAP MAPMAKER	34/34	NI
8*	Donis-Keller (23)	111	171	17	47	3	CEPH (23)	CRI-MAP MAPMAKER	8/10	NI
9*	Donis-Keller (23)	51	110	14	41	1	CEPH (23)	CRI-MAP MAPMAKER	5/8	NI
9	Lathrop (32)	147	231	14	41	4	CEPH (40) Non-CEPH (18)	GMS LINKAGE	10/16	9/16
10	Lathrop (19)	116	170	19	46	1	CEPH (40) Non-CEPH (19)	GMS LINKAGE	9/9	9/9
10	Farrer (28)	92	92	18	24	0	Non-CEPH (10)	LIPED LINKAGE	4/4	4/4
10*	Donis-Keller (23)	94	111	22	50	3	CEPH (23)	CRI-MAP MAPMAKER	3/6	NI
11*	Donis-Keller (23)	166	210	16	45	2	CEPH (23)	CRI-MAP MAPMAKER	10/16	NI
12	O'Connell (33)	111	258	20	92	4	CEPH (28) Non-CEPH (10)	LINKAGE	NI	NI
12*	Donis-Keller (23)	45	97	14	27	1	CEPH (23)	CRI-MAP MAPMAKER	6/8	NI
13*	Donis-Keller (23)	28	79	16	28	1	CEPH (23)	CRI-MAP MAPMAKER	3/6	NI
13	Bowcock (26)	51	78	9	27	1	CEPH (40) Non-CEPH (19)	LIPED LINKAGE	8/9	7/9
13	Haines (27)	37	28	6	10	0	Non-CEPH (18)	LINKAGE	5/5	4/5
13	Leppert (49)	71	267	53	137	3	CEPH (17) Non-CEPH (13)	LINKAGE	NI	NI
14	Nakamura (34)	66	53	7	22	0	CEPH (40) Non-CEPH (19)	GMS LINKAGE	9/10	7/10
15	Nakamura (35)	146	187	16	39	3	CEPH (40) Non-CEPH (19)	GMS	11/13	10/13
15*	Donis-Keller (23)	35	76	10	28	1	CEPH (23)	CRI-MAP MAPMAKER	4/8	NI

Table 3 (Continued)

Chromosome	Author (ref.)	Map length Male	Map length Female	Average distance between loci	Largest gap	No. of gaps > 25 cm	Pedigrees (no. of families)[a]	Programs[b]	Support for locus order[c] > 100:1	Support for locus order[c] > 1000:1
16*	Donis-Keller (23)	164	237	9	45	1	CEPH (23)	CRI-MAP MAPMAKER	25/30	NI
17*	Donis-Keller (23)	26	116	29	50	2	CEPH (23)	CRI-MAP MAPMAKER	3/4	NI
17	Nakamura (36)	218	279	17	51	4	CEPH (40) Non-CEPH (19)	GMS LINKAGE	17/19	13/19
18	O'Connell (55)	97	205	19	57	3	CEPH (40) Non-CEPH (19)	GMS LINKAGE	8/11	8/11
18*	Donis-Keller (23)	81	210	30	50	5	CEPH (23)	CRI-MAP MAPMAKER	6/7	NI
19	Nakamura (37)	137	189	19	46	3	CEPH (40) Non-CEPH (19)	GMS LINKAGE	7/10	6/10
20	Donis-Keller (23)	52	161	32	50	3	CEPH (23)	CRI-MAP MAPMAKER	5/5	NI
21	Tanzi (38)	50	126	14	31	2	Non-CEPH (18)	LIPED LINKAGE	7/9	6/9
21	Warren (39)	114	160	11	26	2	CEPH (40)	LINKAGE MAPMAKER	12/15	10/15
21*	Donis-Keller (23)	41	73	18	30	1	CEPH (23)	CRI-MAP MAPMAKER	4/4	NI
22	Rouleau (58)	100	99	8	20	0	Non-CEPH (18)	LIPED LINKAGE	NI	9/12
22	Julier (30)	71	combined sex	8	32	2	CEPH (12) Non-CEPH (9)	LINKAGE	1/4	1/4
X	Donis-Keller (23)		193	19	66	3	CEPH (23)	CRI-MAP MAPMAKER	9/10	NI
X	Drayna (60)		185	14	77	2	NI	LINKAGE	NI	NI
X	Arveiler (61)		47	7	21	0	CEPH (17) Non-CEPH (27)	LINKAGE	NI	NI

Map length, distance between loci, and gaps are reported in centimorgans calculated using the Haldane mapping function, except for Chromosomes with an asterisk where Kosambi's mapping function was used. The average distance between loci and gap distance was calculated from the reported female genetic maps. The sizes of the gaps for maps reported by Donis-Keller (23) and Murray (24) were truncated at 50 cM and are likely to be larger.

[a]The Centre d'Étude du Polymorphisme Humain (CEPH) is an international collaboration that coordinates the collection and distribution of DNA from documented human pedigrees for linkage analysis. The total number of CEPH families available for analysis is 40; however, this will soon be expanded to 60. Shown in parentheses is the number of families tested for the polymorphic markers listed. Non-CEPH families are families studied for genetic disease linkage or are part of reference pedigrees not included in CEPH.

[b]Computer programs used for constructing linkage maps are generally available from the authors LINKAGE (17), GMS (19), CRI-MAP (18), MAPMAKER (18), LIPED (16).

[c]Most reported genetic maps indicate the support for marker order by calculating the odds against inversion of adjacent marker loci. Shown is the number of adjacent marker pairs where the reported marker order exceeds odds of 100:1 or 1000:1 against inversion of the pair. The maps reported by Donis-Keller (23) and Barker (14) show markers assigned to specific map locations if the odds for a specific location exceed 100:1 over alternative locations but do not specify the statistical support for marker order. For some reported maps, the support for locus order was determined by methods other than calculating the odds against inversion of adjacent marker loci and is therefore listed as not indicated (NI).

are based on multipoint analytical methods, with the exception of the map of chromosome 4 by Murray et al. (24), which is based on pairwise analysis.

Of the 589 markers used to make the linkage maps, 568 are DNA markers that detect RFLPs. Maps that include VNTR markers generally display a higher average frequency of heterozygosity than do maps consisting exclusively of simple restriction site polymorphisms. Although the markers comprising the maps reported by Donis-Keller et al. (23) were not identified as VNTR markers, many of these markers are highly polymorphic and will likely be found to encode VNTR sequences.

Of the 440 genes that are associated with RFLPs (HHMI Human Gene Mapping Library, September 1989), 74 have been assigned to the published chromosome linkage maps (Table 1). Cloned genes that detect RFLPs are particularly useful as genetic markers. Polymorphic DNA probes for genes that are candidates for involvement in genetic disease can be easily tested in family studies. Any evidence of recombination between the gene marker and the genetic disease generally rules out the candidate gene for a role in the disease.

The identification of important genetic disease linkage to DNA markers and subsequent chromosome identification have spurred interest in the intensive mapping of several chromosomes. Two chromosomes with numerous markers and detailed genetic maps are chromosome 7, containing the gene for cystic fibrosis, and chromosome 16, which is the location of the gene for adult polycystic kidney disease. Most of the human chromosomes have now been found to harbor a significant genetic disease gene (Table 1). The genes responsible for cystic fibrosis (CF) (25), Wilson disease (WND) (26), retinoblastoma (RB1) (26,27), multiple endocrine neoplasia type 2A (MEN2A) (28), and hereditary cutaneous malignant melanoma (CMM) (29) have been mapped to specific locations on reported genetic linkage maps by combining data generated from families segregating the genetic disease and from reference pedigrees.

Although many groups have focused their genetic mapping efforts on specific chromosome regions containing genetic disease genes, others have set out to develop foundation maps of chromosomes as part of a larger strategy to map the human genome. This is an ongoing process, and it is expected that the human genome map will evolve in complexity and precision as additional markers are tested and newer families are added to the data base. The goal is to have maps of all the chromosomes with highly polymorphic markers spaced at intervals of about 20 cM. When this goal is reached, genetic disease linkage studies will be significantly enhanced and complex genetic traits that may play a role in cancer and cardiovascular disease can be studied.

The currently status of the human genome map is summarized in Table 3. With one exception [one of the chromosome 22 maps (30)]; all of the published chromosome maps have been evaluated for sex-specific recombination rates. One of the most obvious features of the linkage maps is the apparent difference in the genetic length of the female and male chromosomes. The genetic length of the female chromosome is generally longer than that of males, and in most cases, this difference has been shown to be statistically significant. By combining the different chromosome maps for male and females, the genetic length of the female autosome map is found to be approximately 90% longer than the male map (23). However, this sex difference in recombination is not uniformly distributed throughout the genome since there are chromosome regions where an excess of male recombination is observed [chromosomes 1 (31), 4 (24), 9 (32), 10 (19), 11 (6,23), 12 (33), 13 (26), 14 (34), 15 (35), 17 (36), 19 (37)]. In many cases, the male excess recombination has been observed in the distal regions of chromosomes.

The marker density and overall length of a chromosome map are important features that indicate the completeness of the map. Most maps, although consisting of a single linkage group of markers, still contain gaps. The size and number of gaps will influence the effectiveness of using the current map to localize unmapped markers or genetic disease genes. In Table 3, distances for gaps greater than 50 cM are crude estimates since the mapping functions become highly imprecise for large recombination fractions. Gap distances were truncated at 50 cM (approximately 38% recombination) for all the maps reported by Donis-Keller et al. (23); the map length reported for these chromosomes was therefore conservatively estimated. Since the genetic length of the female chromosome is generally larger than the length of the male chromosome, the density of the genetic maps (average distance between loci) and gap distance were based on the reported female chromosome length. This gives a more accurate picture of the completeness of the map. As noted earlier, two mapping functions (Haldane's and Kosambi's) were used by the different groups when calculating and reporting genetic distance in centimorgans. The differences in overall genetic length resulting from using one mapping function instead of another will be small, therefore, the maps can be directly compared.

Although distances between markers on a map will become more accurate as additional individuals are tested and added to the linkage analysis, once the marker order is defined it should not vary provided markers are placed on the map with an appropriate degree of statistical support. Markers that are closely spaced are often more difficult to resolve since there will be fewer opportunities for recombination. For the published linkage maps included in Table 3, the degree of statistical support for the reported marker interval order is indicated at levels of 100:1 or 1000:1 odds in favor of the reported order over alternative orders. This indicates the general level of confidence in the reported marker order for a specified chromosome map. In many cases, the odds in favor of the reported marker order for an interval defined by a pair of markers greatly exceeds 1000:1.

It is interesting to compare genetic linkage maps based on meiotic recombination to maps made by physically localizing DNA markers by hybridization. One chromosome for which detailed genetic and physical maps exist is chromosome 21. Figure 1 shows the alignment of the genetic linkage maps for this chromosome (38,39) with the physical map (40,41). Twenty-one DNA markers were used by two groups to construct the linkage map; 10 markers are found on both the published maps. The locus order is the same for shared loci, although one group did not detect any recombination between D21S8/APP/D21S111 while the other group was able to order these loci. The reported distances between loci are generally comparable. Differences reflect a different number of meioses used to make the maps since one map is based on analysis of a Venezuelan reference pedigree (38) while the other used the CEPH families (39). The order of the markers determined from the genetic maps is in complete agreement with the marker assignment by physical mapping methods. By comparing the physical and genetic maps, a dramatic difference in the rate of recombination along the chromosome is apparent. Crossovers are concentrated in the terminal region of the chromosome, resulting in over 50% of the genetic map length

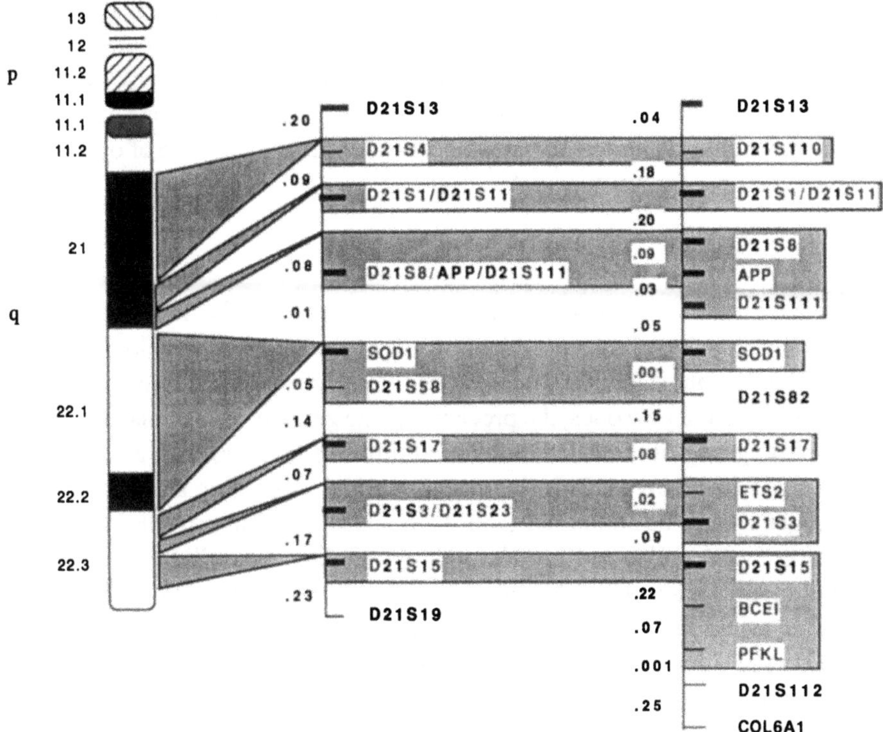

Figure 1 Composite RFLP map of human chromosome 21. Female genetic linkage maps of human chromosome 21 from Warren et al. (39) (right) and Tanzi et al. (38) (left) are compared. Thick horizontal bars show markers shared by the two genetic maps. The recombination fraction for each marker interval is indicated to the left of each map. Markers in shaded regions on the genetic maps are shown against the physical map of chromosome 21 (ideogram of chromosome 21 on the far left). Specific regions on the chromosome are defined by breakpoints of translocation chromosomes contained in a panel of somatic cell hybrids [see Van Keuren et al. (40), Gardiner et al. (41)]. The location of the BCEI locus has been reassigned, as shown, to a position proximal to COL6A1 (S. Antonarakis, personal communication).

localized to about 15% of the chromosome's physical length (band 22.3). This observation is not unique to chromosome 21 as increased recombination in the distal region of chromosome 14 (34) has also been observed. As more complete maps for all the chromosomes are established, including markers for chromosome telomeres, this type of observation can be generally tested.

SUMMARY

Current progress toward obtaining detailed maps of all the human chromosomes has largely been a direct result of the efficient development and application of RFLPs. As more families are studied with the existing markers and new ones, the precision of the genetic linkage maps will increase, gaps will be filled, and ambiguities of marker order and distance will be resolved. This, in turn, will allow interesting areas in human genetics to be addressed, such as the differences in recombination rates between males and females and the nonlinear distribution of recombination rates along the chromosomes. A more immediate impact will be found in the area of medical genetics as markers and maps are used to more efficiently detect the location of genes causing genetic disease and to analyze complex genetic traits. Finally, detailed RFLP maps will provide the starting point for eventually obtaining the nucleotide base sequence of the human genome and further the understanding of human biology and genetics.

REFERENCES

1. Botstein D, White RL, Skolnick M, Davis RW. Construction of a genetic linkage map in man using restriction fragment length polymorphism. Am J Hum Genet 1980; 32:314–331.
2. Gusella JF, Wexler NS, Conneally PM, Naylor SL, Anderson MA, Tanzi RE, Watkins PC, Ottina K, Wallace MR, Sakaguchi AY, Young AB, Shoulson I, Bonilla E, Martin JB. A polymorphic DNA marker genetically linked to Huntington's disease. Nature 1983; 306:234–238.
3. Tsui L-C, Buchwald M, Barker D, Braman JC, Knowlton R, Schumm JW, Eiberg H, Mohr J, Kennedy D, Plavsic N, Zsiga M, Markiewicz D, Akots G, Brown V, Helms C, Gravius T, Parker C, Rediker K, Donis-Keller H. Cystic fibrosis locus defined by a genetically linked polymorphic DNA marker. Science 1985; 230:1054–1057.

4. Wainwright BJ, Scambler PJ, Schmidtke J, Watson EA, Law HY, Farrall M, Cooke HJ, Eiberg H, Williamson R. Localization of cystic fibrosis locus to human chromosome 7cen-q22. Nature 1985; 318:384-386.
5. White R, Woodward S, Leppert M, O'Connell P, Hoff M, Gerbst J, Lalouel J-M, Dean M, Vande Woude G. A closely linked genetic marker for cystic fibrosis. Nature 1985; 318:382-384.
6. White R, Leppert MD, Bishop T, Barker D, Berkowitz J, Brown C, Callahan P, Holm T, Jerominski L. Construction of linkage maps with DNA markers for human chromosomes. Nature 1985; 313:101-105.
7. Wyman A, White R. A highly polymorphic locus in human DNA. Proc Natl Acad Sci USA 1980; 77:6754-6758.
8. Nakamura Y, Leppert M, O'Connell P, Wolff R, Holm T, Culver M, Martin C, Fujimoto E, Hoff M, Kumlin E, White R. Variable number of tandem repeat (VNTR) markers for human gene mapping. Science 1987; 235:1616-1622.
9. Watkins PC, Tanzi RE, Gibbons KT, Tricoli JV, Landes G, Eddy R, Shows TB, Gusella JF. Isolation of polymorphic DNA segments from human chromosome 21. Nucl Acids Res 1985; 13:6075-6088.
10. Litt M, White R. A highly polymorphic locus in human DNA revealed by cosmid-derived probes. Proc Natl Acad Sci USA 1985; 82:6202-6210.
11. Deaven LL, Van Dilla MA, Bartholdi MF, Carrano AV, Cram LS, Fuscoe JC, Gray JW, Hildebrand CE, Moyzis RK, Perlman J. Construction of human chromosome-specific DNA libraries from flow-sorted chromosomes. Cold Spring Harbor Symp Quant Biol 1986; LI:159-167.
12. Fuscoe JC, Clark LM, Van Dilla MA. Construction of fifteen human chromosome-specific DNA libraries from flow-purified chromosomes. Cytogenet Cell Genet 1986; 43:79-86.
13. Gray JW, Carrano AV, Steinmetz LL, Van Dilla MA, Moore DH, Mayall BH, Mendelsohn ML. Chromosome measurement and sorting by flow systems. Proc Natl Acad Sci USA 1975; 72:1231.
14. Barker D, Green P, Knowlton R, Schumm J, Lander E, Oliphant A, Willard H, Akots G, brown V, Gravius T, Helms C, Nelson C, Parker C, Rediker K, Rising M, Watt D, Weiffenbach B, Donis-Keller H. Genetic linkage map of human chromosome 7 with 63 DNA markers. Proc Natl Acad Sci USA 1987; 84:8006-8010.
15. Ott J. Analysis of human genetic linkage. Baltimore: John Hopkins University Press, 1985.
16. Ott J. Estimation of the recombination fraction in human pedigrees: efficient computation of the likelihood for human linkage studies. Am J Hum Genet 1974; 26:588-597.

17. Lathrop GM, Lalouel J-M, Julier C, Ott J. Multilocus linkage analysis in humans: detection of linkage and estimation of recombination. Am J Hum Genet 1985; 37:482–498.
18. Lander ES, Green P, Abrahamson J, Barlow A, Daly M, Lincoln S, Newburg L. MAPMAKER: an interactive computer package for constructing primary genetic linkage maps of experimental and natural populations. Genomics 1987; 1:174–188.
19. Lathrop M, Nakamura Y, Cartwright P, O'Connell P, Leppert M, Jones C, Tateishi H, Bragg T, Lalouel J-M, White R. A primary genetic map of markers for human chromosome 10. Genomics 1988; 2:157–164.
20. Therman E. Human chromosomes: structure, behavior, effects. New York: Springer-Verlag, 1986.
21. Yunis JJ. High resolution of human chromosomes. Science 1976; 191:1268–1271.
22. Franke U, Oliver N. Quantitative analysis of high-resolution trypsin-giemsa bands on human prometaphase chromosomes. Hum Genet 1978; 45:137–165.
23. Donis-Keller H, Green P, Helms C, Cartinhour S, Weiffenbach B, Stephens K, Keith TP, Bowden DW, Smith DR, Lander ES, Botstein D, Akots G, Rediker S, Gravius T, Brown VA, Rising MB, Parker C, Powers JA, Watt DE, Kauffman ER, Bricker A, Phipps P, Muller-Kahle H, Fulton TR, Ng S, Schumm JW, Braman JC, Knowlton RG, Barker DF, Crooks SM, Lincoln SE, Daly MJ, Abrahamson J. A genetic linkage map of the human genome. Cell 1987; 51:319–337.
24. Murray JC, Buetow KH, Smith M, Carlock L, Chakravarti A, Ferrell RF, Gedamu L, Gilliam C, Shiang R, DeHaven CR. Pairwise linkage analysis of 11 loci on human chromosome 4. Am J Hum Genet 1988; 42:490–497.
25. Lathrop GM, Farrall M, O'Connell P, Wainwright B, Leppert M, Nakamura Y, Lench N, Kruyer H, Dean M, Vande Woude G, Lalouel J-M, Williamson R, White R. Refined linkage map of chromosome 7 in the region of the cystic fibrosis gene. Am J Hum Genet 1988; 42:38–44.
26. Bowcock AM, Farrer LA, Hebert JM, Agger M, Sternlieb I, Scheinberg IH, Buys CHCM, Scheffer H, Frydman M, Chajek-Saul T, Bonne-Tamir B, Cavalli-Sforza LL. Eight closely linked loci place the Wilson disease locus within 13q14–q21. Am J Hum Genet 1988; 43:664-674.
27. Haines JL, Ozelius L, St George-Hyslop P, Wexler NS, Gusella JF, Conneally PM. Partial linkage map of chromosome 13q in the region of the Wilson disease and retinoblastoma genes. Genet Epidemiol 1988; 5:375-380.

28. Farrer LA, Castiglione CM, Kidd JR, Myers S, Carson N, Simpson NE, Kidd KK. A linkage group of five DNA markers on human chromosome 10. Genomics 1988; 3:72–77.
29. Bale SJ, Dracopoli NC, Tucker MA, Clark WH, Fraser MC, Stanger BZ, Green P, Donis-Keller H, Housman DE, Greene MH. Mapping the gene for hereditary cutaneous malignant melanoma-dysplastic nevus to chromosome 1p. N Engl J Med 1989; 320:1367–1372.
30. Julier C, Lathrop GM, Reghis A, Szajnert M-F, Lalouel J-M, Kaplan JC. A linkage and physical map of chromosome 22, and some application to gene mapping. Am J Hum Genet 1988; 42:297–308.
31. O'Connell P, Lathrop GM, Nakamura Y, Leppert ML, Ardinger RH, Murray JL, Lalouel J-M, White R. Twenty-eight loci form a continuous linkage map of markers for human chromosome 1. Genomics 1989; 4:12–20.
32. Lathrop M, Nakamura Y, O'Connell P, Leppert M, Woodward S, Lalouel J-M, White R. A mapped set of genetic markers for human chromosome 9. Genomics 1988; 3:361–366.
33. O'Connell P, Lathrop GM, Law M, Leppert M, Nakamura Y, Hoff M, Kumlin E, Thomas W, Elsner T, Ballard L, Goodman P, Azen E, Sadler JE, Cai GY, Lalouel JM, White R. A primary genetic linkage map for human chromosome 12. Genomics 1987; 1:93–102.
34. Nakamura Y, Lathrop M, O'Connell P, Leppert M, Kamboh MI, Lalouel JM, White R. Frequent recombination is observed in the distal end of the long arm of chromosome 14. Genomics 1989 4:76–81.
35. Nakamura Y, Lathrop M, O'Connell, Leppert M, Lalouel J-M, White R. A mapped set of DNA markers for human chromosome 15. Genomics 1988; 3:342–346.
36. Nakamura Y, Lathrop M, O'Connell P, Leppert M, Barker D, Wright E, Skolnick M, Kondoleon S, Litt M, Lalouel J-M, White R. A mapped set of DNA markers for human chromosome 17. Genomics 1988; 2:302–309.
37. Nakamura Y, Lathrop M, O'Connell P, Leppert M, Lalouel J-M, White R. A primary map of ten DNA markers and two serological markers for human chromosome 19. Genomics 1988; 3:67–71.
38. Tanzi RE, Haines JL, Watkins PC, Stewart GD, Wallace MR, Hallewell R, Wong C, Wexler NS, Conneally PM, Gusella JF. Genetic linkage map of human chromosome 21. Genomics 1988; 3:129–136.
39. Warren AC, Slaugenhaupt SA, Lewis JG, Chakravarti A, Antonarakis SE. A genetic linkage map of 17 markers on human chromosome 21. Genomics 1989; 4:579–591.

40. Van Keuren M, Watkins P, Drabkin H, Jabs E, Gusella J, Patterson D. Regional localization of DNA sequences on chromosome 21 using somatic cell hybrids. Am J Hum Genet 1986; 38:793-804.
41. Gardiner K, Watkins P, Munke M, Drabkin H, Jones, Patterson D. Partial physical map of human chromosome 21. Somatic Cell Molec Genet 1988; 14:623-638.
42. Dracopoli NC, Stanger BZ, Ito CY, Call KM, Lincoln SE, Lander ES, Housman D. A genetic linkage map of 27 loci from PND to FY on the short arm of human chromosome 1. Am J Hum Genet 1988; 43:462-470.
43. Bodmer WF, Bailey CJ, Bodmer J, Bussey HJR, Ellis A, Gorman P, Lucibello FC, Murday VA, Rider SH, Scambler P, Sheer D, Solomon E, Spurr NK. Localization of the gene for familial adenomatous polyposis in chromosome 5. Nature 1987; 328:614-616.
44. Leach R, DeMars R, Hasstedt S, White R. Construction of a map of the short arm of human chromosome 6. Proc Natl Acad Sci USA 1986; 83:3909-3913.
45. Mathew CGP, Chin KS, Easton DF, Thorpe K, Carter C, Liou GI, Fong S-L, Bridges CDB, Haak H, Kruseman ACN, Schifter S, Hansen HH, Telenius H, Telenius-Berg M, Ponder BAJ. A linked genetic marker for multiple endocrine neoplasia type 2a on chromosome 10. Nature 1987; 328:527-528.
46. Simpson NE, Kidd KK, Goodfellow PJ, McDermid H, Myers S, Kidd JR, Jackson CE, Duncan AMV, Farrer LA, Brasch K, Castiglione C, Genel M, Gertner J, Greenberg CR, Gusella JF, Holder JJA, White BN. Assignment of multiple endocrine neoplasia type 2a to chromosome 10 by linkage. Nature 1987; 328:528-530.
47. Egeland JA, Gerhard DS, Pauls DL, Sussex JN, Kidd KK, Allen CR, Hostetter AM, Housman DE. Bipolar affective disorders linked to DNA markers on chromosome 11. Nature 1987; 325:783-787.
48. Larsson C, Skogseid B, Oberg K, Nakamura Y, Nordenskjold MC. Multiple endocrine neoplasia type I gene maps to chromosome 11 and is lost in insulinoma. Nature 1988; 332:85-87.
49. Leppert M, Cavenee W, Callahan P, Holm T, O'Connell P, Thompson K, Lathrop GM, Lalouel J-M, White R. A primary genetic map of chromosome 13q. Am J Hum Genet 1986; 39:425-437.
50. Sparkes RS, Murphree AL, Lingua RW, Sparkes MC, Field LL, Funderburk SJ, Benedict WF. Gene or hereditary retinoblastoma assigned to human chromosome 13 by linkage to esterase D. Science 1983; 219:971-973.

51. Frydman M, Bonne-Tamir B, Farrer LA, Conneally PM, Magazanik A, Ashbel A, Goldwitch Z. Assignment of the gene for Wilson disease to chromosome 13: linkage to the esterase D locus. Proc Natl Acad Sci USA 1985; 82:1819–1821.

52. Reeders ST, Breuning MH, Davies KE, Nicholls RD, Jarman AP, Higgs DR, Pearson PL, Weatherall DJ. A highly polymorphic DNA marker linked to adult polycystic kidney disease on chromosome 16. Nature 1985; 317:542–544.

53. Barker D, Wright E, Nguyen K, Cannon L, Fain P, Goldgar D, Bishop DT, Carey J, Baty B, Kivlin J, Willard H, Waye JS, Greig G, Leinwand L, Nakamura Y, O'Connell P, Leppert M, Lalouel J-M, White R, Skolnick M. Gene for von Recklinghausen neurofibromatosis is in the pericentromeric region of chromosome 17. Science 1987; 236: 1100–1102.

54. Seizinger BR, Rouleau GA, Ozelius LJ, Lane AH, Faryniarz AG, Chao MV, Huson S, Korf BR, Parry DM, Periacak-Vance MA, Collins FS, Hobbs WJ, Falcone BG, Iannazzi JA, Roy JC, St George-Hyslop PH, Tanzi RE, Bothwell MA, Upadhyaya M, Harper P, Goldstein AE, Hoover DL, BAder JL, Spence MA, Mulvilhill JJ, Aylsworth AS, Vance JM, Rossenwasser GOD, Gaskell PC, Roses AD, Martuza RL, Breakefield XO, Gusella JF. Genetic linkage of von Recklinghausen neurofibromatosis to the nerve growth factor receptor gene. Cell 1987; 49:589–594.

55. O'Connell P, Lathrop GM, Leppert M, Nakamura Y, Muller U, Lalouel J-M, White R. Twelve loci form a continuous linkage map for human chromosome 18. Genomics 1988; 3:367–372.

56. Bartlett R, Pericak-Vance M, Yamaoka L, Gilbert J, Herbstreith M, Hung W-Y, Lee J, Mohandas T, Bruns G, Laberge C, Thibault M-C, Ross D, Roses A. A new probe for the diagnosis of myotonic muscular dystrophy. Science 1987; 235:1648–1650.

57. St George-Hyslop PH, Tanzi RE, Polinsky RJ, Haines JL, Nee L, Watkins PC, Myers RH, Feldman RG, Pollen D, Drachman D, Growdon J, Bruni A, Foncin J-F, Salmon D, Frommelt P, Amaducci L, Sorbi S, Piacentini S, Stewart GD, Hobbs WJ, Conneally PM and Gusella JF. The genetic defect causing familial Alzheimer's disease maps on chromosome 21. Science 1987; 235:885–890.

58. Rouleau GA, Haines JL, Bazanowski A, Colella-Crowley A, Trofatter JA, Wexler NS, Conneally PM, Gusella JF. A genetic linkage map of the long arm of human chromosome 22. Genomics 1989; 4:1–6.

59. Rouleau GA, Wertelecki W, Haines JL, Hobbs WJ, Trofatter JA, Seizinger BR, Martuza RL, Supermeau DW, Conneally PM, Gusella JF. Genetic linkage of bilateral acoustic neurofibromatosis to a DNA marker on chromosome 22. Nature 1987; 329:246–248.
60. Drayna D, White R. The genetic linkage map of the human X chromosome. Science 1985; 230:753–758.
61. Arveiler B, Oberle I, Mandel JL. Genetic mapping of nine DNA markers in the q11–q22 region of the human X chromosome. Genomics 1987; 1:60–66.
62. Davies KE, Pearson PL, Harper PS, Murray JM, O'Brien T, Sarfarazi M, Williamson R. Linkage analysis of two cloned sequences flanking the Duchenne muscular dystrophy locus on the short arm of the human X chromosome. Nucl Acids Res 1983; 11:2303–2312.

12

The Family of Bovine Prolactin-Related Genes Transcribed in the Placenta

Mark A. Kessler and Linda A. Schuler

University of Wisconsin—Madison
Madison, Wisconsin

INTRODUCTION

Theoretically, any genomic DNA sequence can be used in a search for polymorphisms associated with the phenotype of interest. Practical considerations suggest, however, the use of probes derived from genes which, through independent lines of evidence, may be associated with a desired genetic trait.

The growth hormone/prolactin gene family is of particular interest in the study of production traits in domestic species. The pituitary hormones of this gene family, growth hormone (GH) and prolactin (PRL), are involved in a variety of processes with economic impact for many domestic animals. Growth hormone, in particular, certainly one of the most studied hormones across species, is important in milk production and growth. As discussed below, little is known of the functions of the placental members of this family. Many changes that take place during pregnancy are natural targets of the prolactin-growth hormone gene family, including growth of the fetus and placenta, development of maternal mammary secretory tissue, and modulation of maternal energy metabolism to spare glucose for the fetus. The low levels of PRL during pregnancy, combined with the lack of receptors for GH in the fetus, focuses interest on the novel placental members of the GH-PRL gene family as candidates, in the fetus, for the roles performed postnatally by pituitary hormones, as well as in the unique metabolic adaptations to pregnancy. The impact of our scientific understanding of these processes is apparent in their contribution to reproductive success, the health of offspring, and efficiency in the beef and dairy industries.

It is our long-term goal to understand the function of these related placental hormones in fetal development and maternal adaptations to pregnancy, the basis for hormone and tissue specificity of these actions, and the regulation of their expression. In the balance of this chapter, we describe the historical context of this work and review the available data regarding the potential actions of GH, PRL, and related hormones with particular attention to reproduction, fetal growth, and lactation. We then summarize our progress to date in defining the spectrum of prolactin-related bovine placental transcripts and their products and the structure and chromosomal localization of the corresponding genes.

ORIGIN OF THE GROWTH HORMONE/PROLACTIN GENE FAMILY

Duplication of an ancestral gene about 400 million years ago is believed to have given rise to the pituitary hormones (1-3). The prolactin and growth hormone genes have been the subject of intensive study for some time, and much is known about their structure, regulation, products, and actions across species. In the rat, regulation of expression of the PRL and GH genes has been extensively investigated, providing an important model for hormonal regulation of gene expression. The genes coding for the bovine pituitary hormones have been characterized (4-6), and some aspects of the hormonal control of transcription of the bPRL gene have been reported (7,8).

Human placental lactogen (hPL) was the first placentally expressed member of this gene family to be identified on the basis of its strong resemblance to hGH (9) and high levels of expression during pregnancy, comprising as much as 5% of the mRNA (10) and 10% of the term placenta's protein synthesis (11). Subsequently, it has been found that this hormone is the product of two related genes expressed in the placenta (12). As its early discovery indicates, both the genetic structure and the hormone itself suggest a close evolutionary relationship to hGH (92% nucleotide sequence homology, 85% protein sequence homology). The function of this placental hormone in the human has received much attention, but remains unclear. Deletions of these genes in association with normal gestation and subsequent lactation have been documented in human pregnancies (13,14).

Nonprimates do not appear to have a GH-like placental product comparable to that described in the human, although growth hormone or prolactin-like activity has been observed in many other mammals (15). Instead, the placentae of these species produce products more closely related to PRL. These nonprimate placental products differ from their evolutionary parent more than the human hormone and are expressed at more moderate levels than hPL. In further contrast to the situation in primates, multiple hormones quite distinct from one another have been identified in our studies of the placental products of the cow, as described here (16–18), and in rodents by Talamantes, Linzer, Duckworth, Friesen, and their colleagues (reviewed in Ref. 19). The functional relationships between these various prolactin-related products across species are far from clear, however.

POTENTIAL ACTIONS OF THESE HORMONES

The actions, target tissues, and mode of action of these placental proteins are not known. By analogy to GH and PRL, as noted above, many processes during pregnancy may be regulated by these hormones. GH and PRL themselves may not play major roles at this time, particularly in the ruminant. The fetus has no receptors for GH until after birth (20), and PRL levels are very low during pregnancy, in both ruminants and rodents (21–22). It seems likely, therefore, that the placental relatives of this gene family may play critical roles during gestation, but this has been difficult to study for several reasons. First, traditional ablation experiments designed to test the effect of supplemental hormone are not possible in this system, since destruction of the placenta is fatal to the fetus. Second, only one of the multiple related placental hormones has been isolated in ruminants, limiting the scope of any studies. Third, homologous peptide hormones may demonstrate only relative specificity of receptor binding, and hence activity, making interpretation difficult. Finally, considerable species variation in structure of both hormones and receptors and the use of heterologous systems have complicated analysis of target tissue response (2,23). In particular, the use of the readily available hPL has confused these studies. The following is a brief discussion of some possible roles these hormones may play during pregnancy.

Fetal Growth

Control of fetal growth is not well understood. Although GH is an important determinant of growth in the postnatal animal, the fetus cannot respond to GH, apparently because of the lack of receptors for this hormone prenatally (20,24). Availability of maternal nutrients, fetal ability to acquire and respond to fluctuations in nutrient levels, and local control mechanisms appear to be very important for fetal growth. Insulin, insulin-like growth factors (IGF-1 and -2), and local tissue growth factors have been implicated in the growth and differentiation of fetal tissues (reviewed in Ref. 25). A biochemically distinct fetal IGF-I receptor that declines rapidly in the first weeks of postnatal life has recently been identified in rat skeletal muscle (26). Control of fetal IGFs is not well understood; it has been suggested that placental hormones may help control IGF levels, particularly IGF- II, similarly to GH action on IGF-I in the postnatal mammal, although evidence remains inconclusive (27–31). These hormones may contribute to regulation of IGF serum binding proteins, which are developmentally regulated (32,33). Binding of mouse proliferin to hepatic IGF-II/mannose-6-phosphate receptors has been demonstrated (34); however, the physiological significance of this observation is not yet established. The synlactin hypothesis (35) proposed the existence of factors secreted from the liver in response to PRL which act in concert with this hormone to achieve its biological effect. This has recently been confirmed by purification of an 85-amino-acid "liver lactogenic factor" (36), which is synergistic with PRL in promoting galactopoiesis in rat mammary cells in vitro. Thus, for the future study of the action of placental members of this gene family, the pituitary hormones provide established models of direct (37,38), mediated and synergistic mechanisms.

Growth of the Placenta

Like the growth of the fetus, the mass of the placenta itself increases exponentially early in gestation, and its size and function determine fetal access to maternal gas exchange and nutrients. Little is known about factors controlling the growth of this important tissue. IGF-I, IGF-II, and their receptors have been identified in human and mouse placentae (39–41). These placental prolactin-related hormones may regulate their synthesis or act directly on this tissue. In this regard, the rat decidual prolactin-like protein has been shown to bind specifically to decidual tissue as well as rat ovary (42,43), and PRL binding sites have been characterized on human chorion (44). Synthesis of one of the mouse placental proteins was originally discovered as a response of Balb/c 3T3 cells to mitogens, suggesting potential for autocrine or paracrine action (45,46).

Maternal Metabolism

These placental hormones may also exert effects on the mother to promote fetal and postnatal welfare of the offspring. They may modulate maternal energy metabolism to increase fetal access to glucose, via GH-like anti-insulin effects on the liver and peripheral tissues (47). In the nonpregnant sheep, placental lactogen (PL) has been shown to promote lipolysis (48). Somatogenic receptors that bind oPL have been shown to increase during pregnancy in the sheep liver (49); lactogenic receptors with greater affinity for mPLII than mPRL increase during pregnancy in the liver of the mouse (50).

Mammogenesis and Lactogenesis

The role of these hormones in mammary development and lactation has received considerable attention (51). Indirect evidence suggests that a placental hormone in this family rather than pituitary PRL stimulates the mammary development that occurs during pregnancy. Suppression of PRL secretion in ewes and heifers had no effect on mammogenesis during pregnancy, and in goats, lactation was normal in bromocryptine-treated goats bearing twins (52,53). It has long been known that the sire of the fetus affects subsequent milk production of the mother (54,55). Placental hormones, such as these prolactin-related proteins, whose production would be controlled by the genotype of the fetus, might explain this. PL in both the sheep and cow, in contrast to the mouse, does not appear to be lactogenic (56–58).

Immune Competence

Again, by analogy to activities of PRL and GH, several other functions suggest themselves. The role of PRL in immune function is becoming apparent (59–61). It is possible that these hormones may contribute to the development of the fetal immune response, or to the modulation of the maternal immune system that occurs during pregnancy. In the mouse, a species in which maturation of the immune system occurs shortly after birth, administration of anti-PRL serum altered the developmental pattern of T and B cells in the spleen and thymus of the neonates (62). Immune competence in other species, including the bovine, develops during gestation, and these placental hormones may play a role similar to murine PRL.

Other Possible Roles

Finally, these hormones may play a role in steroidogenesis, of either the placenta or corpus luteum (63). In addition, they may regulate the osmotic balance across the skin of the fetus and amnion (64).

NONPRIMATE PLACENTAL HORMONES

Placental Lactogens

Despite the importance of the potential functions of these placental members of the prolactin–growth hormone gene family, their identification has progressed slowly. Their relationship to the pituitary hormones has proven useful, as they were originally identified in many species on the basis of their ability to interact with PRL or GH receptors (15,22). For these studies, as well as for many of the assay procedures used to detect the proteins during purification, rabbit mammary gland tissue was used as a source of PRL-preferring, or lactogenic receptor, because of its affinity for PRLs of many species. Pregnant rat or rabbit liver has classically been used as GH, or somatogenic receptor. Using these assays, two placental and one decidual protein have been detected in rodents (42,65–67). In other species, including ruminants, this methodology has resulted in the isolation of a single protein (68–73). Here we will call these placental members of the prolactin-growth hormone gene family isolated by radioreceptor assay placental lactogens (PL), consistent with the literature.

Bovine Prolactin-Related Transcripts

Identification Using Low-Stringency Hybridization

In order to identify additional members of this gene family, work in this laboratory also exploited the relationship to the pituitary hormones. However, instead of relying on ability to cross-react with one another's receptors, we looked for homology in their transcripts. Using low-stringency hybridization, we have identified genes related to the growth hormone–prolactin gene family expressed in the placenta which were not detected by the classical receptor assays. Screening of a cDNA library with bovine PRL and GH cDNAs at low stringency revealed a family of cDNAs (at least five discrete transcripts) that are more similar to PRL than GH (16–18). The predicted proteins are only about 40–50% similar to bPRL and about 20% to bGH. They are related to one another by 60–89% at the level of nucleotide sequence and 30–75% at the level of predicted amino acid sequence. bPL is quite distinct from the remaining cDNAs. It is more closely related to PRL and only 30% similar in predicted amino acid sequence to the other cDNAs, which comprise a diverse subfamily (45–75% amino acid similarity to each other). As shown in Figure 1, the differences in amino acid sequence between these predicted placental hormones and PRL are spread throughout the length of the coding sequence. Our laboratory has focused on bPL and bovine prolactin-related cDNA I (bPRCI) as an example for the other subgroup for many of our studies.

Levels of mRNAs Throughout Gestation

In the bovine placenta, steady-state levels of the mRNAs represented by the placental prolactin-related cDNAs are moderately high but much less than those coding for hPL [0.05–0.40% of the poly (A)+ RNA, varying with the mRNA species, at 6 months' gestation]. mRNA levels corresponding to these genes vary over the course of gestation, peaking in the latter third of pregnancy.

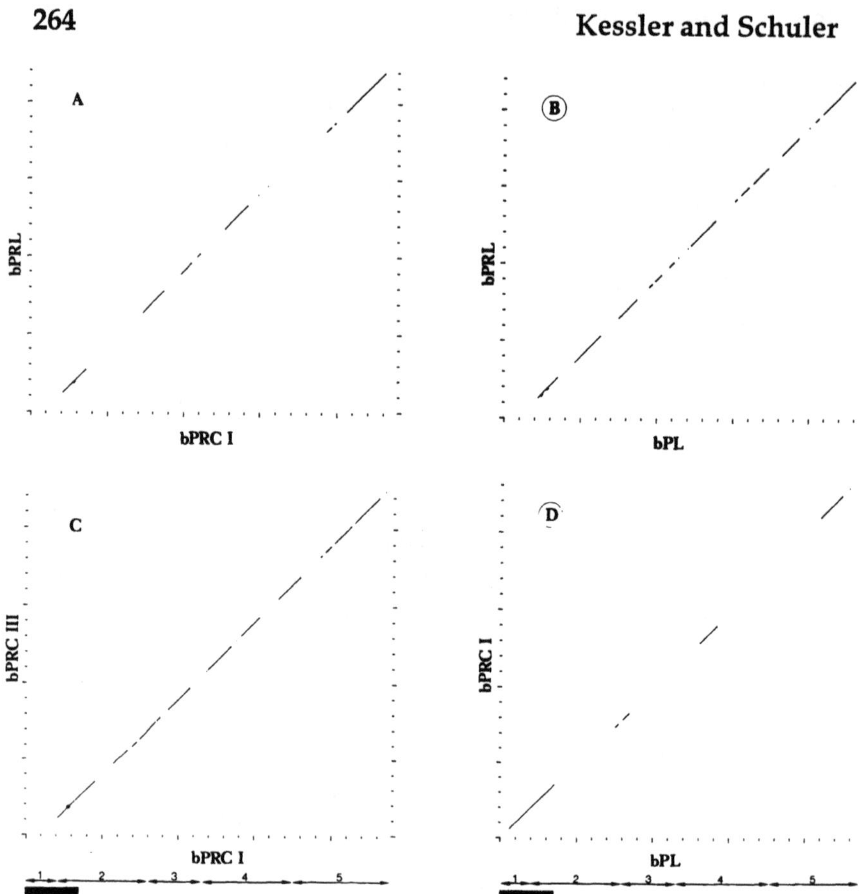

Figure 1 Comparison of the amino acid sequences of some of these placental prolactin-related mRNAs to one another and to bPRL using the computer program, Dotplot (116). 100% homology would be indicated by a solid diagonal line. Exons encoding the amino acids are diagrammed below. The solid rectangle shows the location of the signal peptide. Amino acid sequences were derived from coding regions of cDNAs (bPRL, 116; bPL, 17; bPRCI, 16; bPRCIII, 18).

Predicted Proteins

The amino termini of all the predicted proteins contain a strongly hydrophobic region homologous to the signal peptide of bovine prolactin. Should this region function similarly for these hormones, most of these cDNAs would predict secreted hormones of about 200 residues, similar to the size of the related pituitary hormone. They are structurally somewhat similar to PRL, indicating potential to form two to three disulfide bonds, depending on the site of putative signal peptide cleavage. All but the smallest predicted proteins contain one to four consensus sites for N-glycosylation, suggesting that they may be glycoproteins, in contrast to most of the pituitary members of this gene family (3). Our data show that this is the case (see below).

Alternative Splicing

Most of the cDNAs we have identified predict prehormones of 27–29 kDa, confirmed by in vitro translation of hybrid-selected mRNA. However, several interesting members code for both larger and smaller products. These unusual cDNAs correspond to naturally present transcripts, which also translate in vitro. Analysis of translation products of mRNA hybrid-selected with bPRCI cDNA using nonequilibrium pH gradient gel electrophoresis (NEPHGE; Fig. 2) shows a family of related products, with the major products at about 27 kDa and more minor products at 18–20 kDa and 43 kDa. Similar analysis of mRNA selected with bPL cDNA shows minor products of about the same size. Smaller products have been described for other members of this gene family, including those formed by alternative splicing (see below) and a 16-kDa proteolytically modified form of rPRL. The latter displays a pattern of target tissues distinct from intact rPRL, suggesting different functions (74).

Based on our analysis of the gene structure of the representative bPL and bPRCI genes (see below), at least some of the smaller related products observed in the bovine placenta are apparently the products of alternatively spliced transcripts. The variant transcripts for the bPL gene include one species without those sequences contained within the third exon and another where part of the fourth exon has been spliced out, resulting in entrance into the coding sequences of the fifth exon in a shifted reading frame (75). Products of alternate splicing have also been

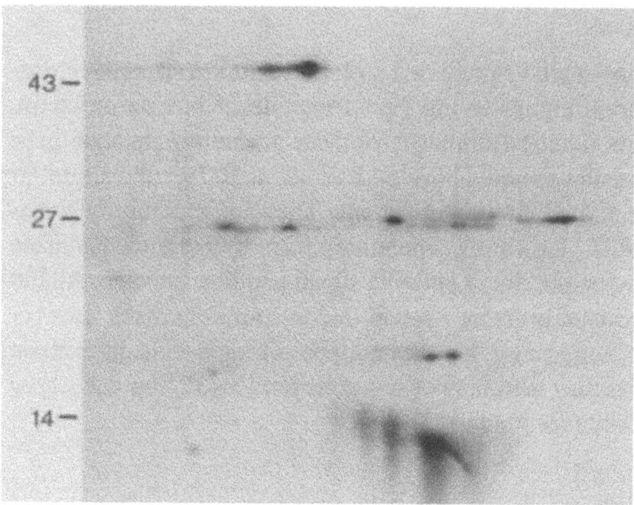

Figure 2 Bovine cotyledonary poly (A)+ RNA from 180d gestation was hybrid-selected with bPRCI in 50% formamide, 410 mM NaCl, 20 mM PIPES, pH 6.4, 1 mM EDTA, 0.2% SDS, 125 ug/ml tRNA, for 2 h at 50°, and 3 h at 43°C. Selected RNA was translated in vitro using a reticulocyte lysate system and products resolved by NEPHGE. Signals at 47 and 14M_r represent artifacts of the system, nonenzymatically labeled protein, and hemoglobin, respectively. (Reprinted with permission from Ref. 18. Copyright 1989 American Chemical Society.)

reported for several other members of this gene family, including the hGH gene, resulting in a 20-kDa GH (76) and a predicted 17.5-kDa product missing the sequences for the third exon, similar to our finding for a variant bPL transcript (77). A failure to remove the final intron has been identified in transcripts of the bGH and hGH-V genes (78,79). Except for the 20-kDa GH protein, the products encoded by these splicing variants have not been characterized. The sequences of the proteins encoded by these novel transcripts predict significant structural differences from the studied hormones, including missing or displaced Cys residues, which form internal disulfide bonds. Further study is necessary to document the products of these variant transcripts and their biological activities.

In Vivo Products

Although the sequence of the cDNAs prepared from these placental transcripts gave no evidence of any structural problems, and translation of the hybrid selected mRNA showed that at least several of these transcripts could be translated in vitro, additional proof was necessary to demonstrate that these genes were in fact expressed in vivo. In order to identify the products of these genes, we prepared a fusion gene of bPRCI with beta-galactosidase, overexpressed the product in *Escherichia coli*, and used the fusion protein to obtain antisera in rabbits (80). This antiserum does not cross-react with either of the related pituitary hormones or bPL, nor does bPL antiserum react with recombinant bPRCI protein. Using this antiserum, we have demonstrated multiple immunoreactive products in the bovine placenta by Western blot of a NEPHGE separation of bovine placental proteins. At this time, the precise relationship between the in vivo protein isoforms detected thus far and the complex pattern of translation products (Fig. 2) is not clear. However, posttranslational modification does apparently contribute to the pattern of observed natural products. For example, one-dimensional Western blot analysis of bovine placenta using our fusion protein antiserum detects a doublet of about 35 kDa. Prior treatment of the sample with endoglycosidase F collapses these two signals into a single band of about 23.5 kDa, the predicted size of the primary translation product. Thus, like bPL (81), the product of bPRCI is posttranslationally modified by N-glycosylation.

Relationship to Rodent Placental Hormones

The differences in the primary structure predicted by these cDNAs from one another and from bovine prolactin suggest that at least one receptor other than the characterized lactogenic and somatogenic receptors mediates the actions of these products. The apparent inability of the product of bPRCI to bind to rabbit PRL and GH receptors is consistent with this. Additional PRL-related genes, such as those we have described in the cow, have not been identified in primates, and other domestic species have not been examined. However, in both the rat and mouse, a story similar to our findings in the bovine has been reported by Duckworth, Friesen, Linzer, Talamantes, and their colleagues, although comparison of the predicted amino acid sequences of the bovine clones to those of rodents shows only about 30% amino acid similarity (82–85). In light of the low level of homology of these predicted hormones between species, our lack of understanding of the domains that interact with the receptor (reviewed in Ref. 86) and hence are responsible for the activity of these hormones, and known species differences in the activity of even the well-characterized members of this gene family, the relationship of individual predicted bovine hormones to the predicted rodent hormones is unclear. Conservation of residues throughout evolution has not been a good prognosticator of function for this gene family. Recent studies of recombinant hGH using in vitro scanning mutagenesis indicate that the overall folding of the protein and binding to the hGH receptor can be maintained despite a surprising number of amino acid substitutions. Those residues which are important for receptor binding are not contiguous in the primary sequence but are in close proximity in the folded structure (87,88). These studies were greatly facilitated by the availability of cloned hGH receptor (89). The availability of cloned rat prolactin receptor (90) should make possible similar studies for prolactin-like hormones.

CHROMOSOMAL ORGANIZATION OF THE GENES FOR THE BOVINE PROLACTIN-RELATED HORMONES

Genomic Southern Analysis

The complexity of this gene family is demonstrated in Figure 3. Hybridization of one of the isolated cDNAs to a filter containing genomic DNA cut with various restriction enzymes under conditions of lowered stringency shows multiple bands. However, when the same blot is hybridized to the same cDNA using more stringent conditions, the pattern is much simpler.

Structural Genes for bPL and bPRCI

In order to begin to study the structure of these genes and provide the basis for studies of their regulation, we have focused on the bPL gene (75) and that corresponding to a member of the related placental subfamily, bPRCI (91). We screened a bovine genomic library in lambda Charon 28 (4) under conditions of high stringency with these cDNAs

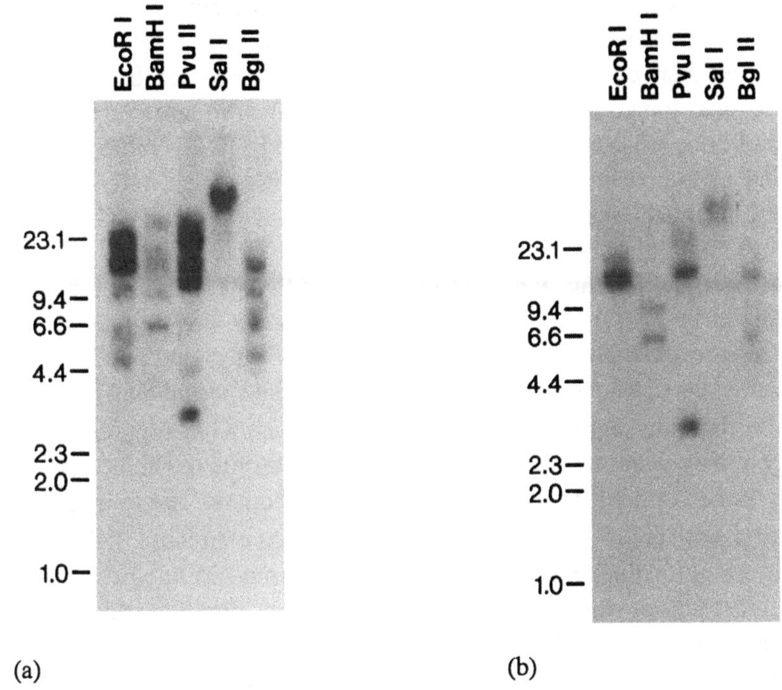

Figure 3 Bovine placental genomic DNA was cut to completion with the restriction enzymes shown, fractionated in 1.0% agarose, transferred to a hybridization filter, and hybridized to bPRCIII at (a) moderate stringency (50% formamide, 5xSSPE, 5x Denhardts, 0.1% SDS, 42°C; washes: 2xSSC, 0.2% SDS, 65°C × 30 min, 0.1xSSC, 0.1% SDS, 65°C 30 min), or (b) high stringency (50% formamide, 1.25xSSPE, 5x Denhardts, 0.1% SDS, 42° washes as above, except that the final wash was continued for a total of 1 h). Molecular size markers are indicated. (Reprinted with permission from Ref. 18. Copyright 1989 American Chemical Society.)

and identified clones containing portions of these chromosomal genes. Restriction enzyme maps of these genes are shown in Figure 4. Like the other members of this gene family, the coding regions are divided into five exons, joined at conserved splice sites (3). The larger structural span of both bPL and the gene corresponding to bPRCI is more similar to that of the prolactin members of the GH-PRL gene family. These genes are much larger (bPRL, 10 kb; bPL, about 12 kb; bPRCI, 9 kb; 6,75,91) than that for growth hormone (2.8 kb; 4,5).

All members of this gene family examined to date, including the pituitary hormones, are homologous throughout the coding regions of the mature hormone and the hydrophobic region of the signal peptide (3). These sequences are encoded by exons 2–5 of the chromosomal genes. The 5′ flanking region and first exon, in contrast, have apparently diverged widely among many family members. Despite the homology of both the bPL and bPRCI genes to bPRL throughout the last four exons, the first exon and 5′ flanking regions of the genes expressed in the placenta are quite distinct from bPRL. However, from 205 bp upstream of the transcription start site through the first exon, the bPL and bPRCI genes share significantly higher homology than they do in the remainder of their coding sequences. The conserved sequences within this region may contribute to the observed similarity in temporal and spatial expression. Both bPL-and bPRCI-specific transcripts are elevated during the last two-thirds of pregnancy (18), and both are expressed in the granulated binucleate cell of the fetal placenta (92).

Both genes have relatively short 5′ untranslated regions, and TATA box-like elements (93) are present in the appropriate positions upstream from the apparent cap sites. Several sequence motifs are present in the 5′ flanking regions of both genes which present interesting possibilities for control of these genes. Both have sequences similar to the phorbol ester response element (94). Phorbol ester modulation of expression has been demonstrated for several other members of this gene family, including rat PRL and mouse proliferin (95,96). Further studies will be necessary to show the functional significance of these sequences in the cell-specific expression and regulation of these genes.

These observations suggest that these placentally expressed genes may have arisen from a common gene duplicated from PRL, which acquired a first exon and 5′ flanking region distinct from the parent gene.

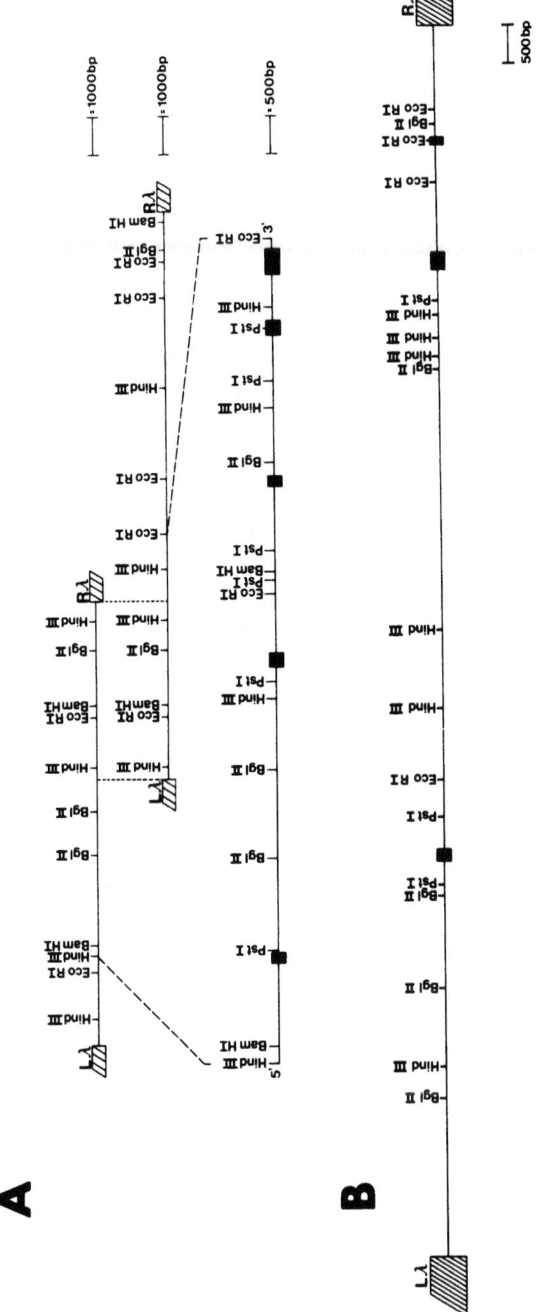

Figure 4 (A) Structural organization of the gene corresponding to bPRCI (modified with permission from Ref. 91). Information from overlapping genomic clones was combined to give the genomic map. (B) Lambda clone containing the 5′ portion of the bPL gene (75). Bovine genomic DNA is depicted as a solid line, and lambda Charon 28 vector as hatched. Exons are indicated by solid boxes.

Subsequent divergence of bPL from the other subfamily, represented by bPRCI, and perhaps evolutionary pressure on bPL to remain similar to PRL or crossingover may have resulted in the modern bovine genes.

Chromosomal Location

The genes coding for the pituitary hormones in nonprimates have diverged widely, as in the human. In primates, where most of the work has been done, the GH gene cluster (including the GH gene expressed in the pituitary and closely related placental homologs) is located on a separate chromosome from PRL. The gene for PRL is located on chromosome 6, whereas the five closely related genes belonging to the GH locus are found on the long arm of human chromosome 17, spanning about 60,000 base pairs (97–99). The high homology among the GH- and PL-related genes in primates and their close physical proximity are consistent with relatively recent duplication events, possibly including subsequent crossingover contributing to concerted evolution (1–3,100–102). Data from the cow, as well as the mouse and rat, indicate that the PRL and GH genes are also located on different chromosomes, demonstrating that the dispersal of these genes probably occurred before the mammalian radiation (103–105).

We have examined the chromosomal location of the genes for the placental members of this gene family, including bPL and bPRCI, by analysis of somatic cell hybrids (106,107). All these placentally expressed genes in the cow are syntenic with the U20 syntenic group containing the bPRL gene, located on chromosome 23 (108). This chromosome has been shown to contain linkage groups also found on human chromosome 6 (108). Linzer and colleagues (104) have shown that the PRL gene and the placentally expressed genes in the mouse (mPLI, mPLII, proliferin, and mouse proliferin-related protein) are located on chromosome 13, which also displays evolutionary conservation with human chromosome 6. The conserved linkage groups of the prolactin-related genes across species are consistent with their closer relationship to PRL than GH and may indicate a more recent evolutionary derivation and/or shared regulatory elements among the syntenic hormonal genes. Future studies will be necessary to determine the location and relative positions of these prolactin-related genes on the bovine chromosome by procedures described elsewhere in this book, which can be used to address these questions.

The syntenic group defined on chromosome 23 in the cow, including PRL itself, the related placental hormones, glyoxalase, steroid 21-hydroxylase, and several genes of the major histocompatibility complex (MHC) make this chromosome one of the most defined in this species (109,110). A homologous region of the short arm of human chromosome 6 includes the MHC, C4 complement, tumor necrosis factor-beta, and hPRL loci (111). The grouping of these genes with important implications for immune function and reproduction traits points to this chromosome as an important target for RFLP studies. Appreciable numbers of RFLPs have been observed around the loci of the genes for PRL and the placental hormones, even within a single strain of cattle (112,113, unpublished observations). One such polymorphism around the bPRL gene has been shown to correlate with total milk production within a major Holstein family (114). Such evidence suggests that these genes may be useful markers for economic traits.

SUMMARY

We have shown how our understanding of gene families can be used to identify previously unknown members and how this "reverse genetics" can be applied to isolate these gene products and begin studies of their function. Our information about the structure of these genes, their chromosomal organization, and regulation of their expression will increase understanding of the bovine genome and the molecular basis for regulation of gene expression in this species. The accumulation of nucleotide sequence will simplify future genetic studies by RFLP analysis, enabling design of hybridization probes capable of distinguishing individual members of the bovine prolactin-related gene family. These parallel genetic and biochemical approaches will aid our understanding of the mechanisms underlying genetic diversity.

ACKNOWLEDGMENTS

This work was supported in part by NSF grants DCB8608739 and DCB8819282 and USDA Grant 88-37240-4103.

REFERENCES

1. Niall DH, Hogan ML, Sauer R, Rosenblum IY, Greenwood FC. Sequence of pituitary and placental lactogenic and growth hormone: evolution from a primordial peptide by gene reduplication. Proc Natl Acad Sci USA 1971; 68:866-869.
2. Wallis M. The molecular evolution of pituitary growth hormone, prolactin and placental lactogen: a protein family showing variable rates of evolution. J Mol Evol 1981; 17:10-18.
3. Miller WL, Eberhardt NL. Structure and evolution of the growth hormone gene family. Endocr Rev 1983; 4:97-130.
4. Woychik RP, Camper SC, Lyons RL, Horowitz S, Goodwin EC, Rottman FM. Cloning and nucleotide sequencing of the bovine growth hormone gene. Nucl Acids Res 1982; 10:7197-7210.
5. Gordon DF, Quick DP, Erwin CR, Donelson JE, Maurer RA. Nucleotide sequence of the bovine growth hormone chromosomal gene. Mol Cell Endocrinol 1983; 33:81-95.
6. Camper SA, Luck DN, Yao Y, Woychik RP, Goodwin RG, Lyons RH, Rottman FM. Characterization of the bovine prolactin gene. DNA 1982; 3:237-249.
7. Camper SA, Yao YAS, Rottman FM. Hormonal regulation of the bovine prolactin promoter in rat pituitary tumor cells. J Biol Chem 1985; 260:12246-12251.
8. Sakai DD, Helms S, Carlstedt-Duke J, Gustafsson J-A, Rottman FM, Yamamoto KR. Hormone-mediated repression: a negative glucocorticoid response element from the bovine prolactin gene. Genes Dev 1988; 2:1144-1154.
9. Josimovich JB, MacLaren JA. Presence in the human placenta and term serum of a highly lactogenic substance immunologically related to pituitary growth hormone. Endocrinology 1962; 77:209- 220.
10. Boime I, McWilliams, D, Szczesna E, Camel M. Synthesis of human placental lactogen messenger RNA as a function of gestation. J Biol Chem 1976; 251:820-825.
11. Kaplan SL, Gurpide D, Sciarra JJ, Grumbach MM. Metabolic clearance rate and production rate of chorionic growth hormone-prolactin in late pregnancy. J Clin Endocrinol Metab 1968; 28:1450- 1460..
12. Barrera-Saldana HA, Seeburg PH, Saunders GF. Two structurally different genes produced the same secreted human placental lactogen hormone. J Biol Chem 1983; 258:3787-3793.

13. Nielsen PA, Pedersen H, Kampmann EM. Absence of human placental lactogen in an otherwise uneventful pregnancy. Obstet Gynecol 1979; 57:332-336.
14. Wurzel JM, Parks JS, Herd JE, Nielsen PV. A gene deletion is responsible for absence of human chorionic somatotropin. DNA 1982; 1:251-257.
15. Talamantes F. A comparative study of the occurrence of placental prolactins among mammals. Gen Comp Endocrinol 1975; 27:115-121.
16. Schuler LA, Hurley WL. Molecular cloning of a prolactin-related mRNA expressed in bovine placenta. Proc Natl Acad Sci USA 1987; 84:5650-5654.
17. Schuler LA, Shimomura K, Kessler MA, Zieler CG, Bremel RD. Bovine placental lactogen: molecular cloning and protein structure. Biochemistry 1988; 27:8443-8448.
18. Kessler MA, Milosavljevic M, Zieler CG, Schuler LA. A subfamily of bovine prolactin-related transcripts distinct from placental lactogen in the fetal placenta. Biochemistry 1989; 28:5154-5161.
19. Ogren L, Talamantes F. Prolactins of pregnancy and their cellular source. Int Rev Cytol 1988; 112:1-65.
20. Gluckman PD, Butler JH, Elliot TB. The ontogeny of somatotropic binding sites in ovine hepatic membranes. Endocrinology 1983; 112:1607-1612.
21. Shiu RPC, Kelly PA, Friesen HG. Radioreceptor assay for prolactin and other lactogenic hormones. Science 1973; 180:968-971.
22. Kelly PA, Tsushima T, Shiu RPC, Friesen HG. Lactogenic and growth hormone-like activities in pregnancy determined by radioreceptor assays. Endocrinology 1976; 99:765-774.
23. Nicoll C. Prolactin and growth hormone: specialists on one hand and mutual mimics on the other. Perspectives Biol Med 1982; 25:369-380.
24. Parkes MJ, Hill DJ. Lack of growth hormone-dependent somatomedins or growth retardation in hypophysectomized fetal lambs. J Endocrinol 1985; 104:193-199.
25. Gluckman PD. Regulation of fetal growth. In: Schuler LA, First NL, eds. Regulation of growth and lactation in animals. Madison: University of Wisconsin Press, 1985:81-84.
26. Alexandrides TK, Smith RJ. A novel fetal insulin-like growth factor (IGF) I receptor: mechanism for increased IGF I-and insulin-stimulated tyrosine kinase activity in fetal muscle. J Biol Chem 1989; 264:12922-12930.

27. Hurley TW, D'Ercole AJ, Handwerger S, Underwood LE, Furlannetto RW, Fellow RE. Ovine placental lactogen induces somatomedin: a possible role in fetal growth. Endocrinology 1977; 101:1635-1638.
28. Adams SO, Nissley SP, Handwerger S, Rechler MM. Developmental pattern of insulin-like growth factor I and II synthesis and regulation in fibroblasts. Nature 1983; 302:150-153.
29. Gluckman PD, Butler JH. Parturition-related changes in insulin-like growth factors I and II in the perinatal lamb. J Endocrinol 1983; 99:223-227.
30. Freemark M, Comer M, Korner G, Handwerger S. A unique placental lactogen receptor: implications for fetal growth. Endocrinology 1987; 120:1865-1872.
31. Mesiano S, Young IR, Baxter RC, Hintz RL, Browne CA, Thorburn GD. Effect of hypophysectomy with and without thyroxine replacement on growth and circulating concentrations of insulin-like growth factors I and II in the fetal lamb. Endocrinology 1987; 120:1821-1830.
32. Hey AW, Browne CA, Thorburn GD. Fetal sheep serum contains a high molecular weight insulin-like growth factor (IGF) binding protein that is acid stable and specific for IGF-II. Endocrinology 1987; 121:1975-1984.
33. Gelato MC, Rutherford C, Stark RI, Daniel SS. The insulin-like growth factor II/mannose-6-phosphate receptor is present in fetal and maternal sheep serum. Endocrinology 1989; 124:2935-2943.
34. Lee S-J, Nathans D. Proliferin secreted by culture cells binds to mannose 6-phosphate receptors. J Biol Chem 1988; 263:3521-3527.
35. Nicoll CS, Herbert NJ, Russell SM. Lactogenic hormones stimulate the liver to secrete a factor that acts synergistically with prolactin to promote growth of the pigeon crop-sac mucosal epithelium in vivo. Endocrinology 1985; 116:1449-1453.
36. Frawley LS, Schwabe C, Miller HA, Betts JG, Hoeffler JP, Simpson MT. Characterization and physiologic role of a liver lactogenic factor. In: Hoshino K, ed. Prolactin gene family and its receptors. Amsterdam: Excerpta Medica, 1988:49-59.
37. Isaksson OGP, Eden S, Jansson J-O. Mode of action of pituitary growth hormone on target cells. Am Rev Physiol 1985; 47:483-499.
38. Kostyo J. Growth hormone: receptors and mode of action. In: Schuler LA, First NL, eds. Regulation of growth and lactation in animals. Madison: University of Wisconsin Press, 1985:35-40.
39. Pilistine SJ, Moses AC, Munro HN. Insulin-like growth factor receptors in rat placental membranes. Endocrinology 1984; 115:1060-1065.

40. Fant M, Munro H, Moses AC. Production of insulin-like growth factor binding protein(s) by human placenta: variation with gestational age. Placenta 1988; 9:397–407.
41. Read LC, Ballard FJ, Francis GL, Baxter RC, Bagley CJ, Wallace JC. Comparative binding of bovine, human and rat insulin-like growth factors to membrane receptors and to antibodies against human insulin-like growth factor-1. Biochem J 1986; 233:215- 221.
42. Jayatilak PD, Glaser LA, Basuray R, Kelly PA, Gibori G. Identification and partial characterization of a prolactin-like hormone produced by rat decidual tissue. Proc Natl Acad Sci USA 1985; 82:217–221.
43. Jayatilak PD, Gibori G. Ontogeny of prolactin receptors in rat decidual tissue: binding by a locally produced prolactin-like hormone. J Endocrinol 1986; 110:115–121.
44. Herington AC, Graham J, Healy DL. The presence of lactogen receptors in human chorion laeve. J Clin Endocrinol Metab 1980; 51:1466–1468.
45. Nilsen-Hamilton M, Shapiro JM, Massoglia SL, Hamilton RT. Selective stimulation by mitogens of incorporation of 35 S-methionine into a family of proteins released into the medium by 3T3 cells. Cell 1980; 20:19–28.
46. Linzer DIH, Nathans D. Nucleotide sequence of a growth-related mRNA encoding a member of the prolactin-growth hormone family. Proc Natl Acad Sci USA 1984; 81:4255–4259.
47. Williams RH. Endocrinology, 7th ed. Philadelphia: Saunders, 1985.
48. Thordarson G, McDowell GH, Forsyth IA, Smith SV. Metabolic effects of placental lactogen in sheep. Proc Nutr Soc Aust 1983; 8:127.
49. Emane MN, Delouis C, Kelly PA, Djiane J. Evolution of prolactin and placental lactogen receptors in ewes during pregnancy and lactation. Endocrinology 1986; 118:695–699.
50. Harigaya T, Smith WC, Talamantes F. Hepatic placental lactogen receptors during pregnancy in the mouse. Endocrinology 1988; 122: 1366–1372.
51. Forsyth IA. The roles of placental lactogen and prolactin in mammogenesis: an overview of in vivo and in vitro studies. In: Hoshino K, ed. Prolactin gene family and its receptors. Amsterdam: Excerpta Medica, 1988:137–153.
52. Schams D, Russe I, Schallenberger E, Prokopp S, Chan JSD. The role of steroid hormones, prolactin and placental lactogen on mammary gland development in ewes and heifers. J Endocrinol 1984; 102:121–130.

53. Forsyth IA, Byatt JC, Iley S. Hormone concentrations, mammary development and milk yield in goats given long-term bromocriptine treatment in pregnancy. J Endocrinol 1985; 104:77–85.
54. Skjervold H, Fimland E. Evidence for a possible influence of the fetus on the milk yield of the dam. Ztschr Tierzuckt Zuchtungsbiol 1975; 92:245–251.
55. Adkinson RW, Wilcox CJ, Thatcher WW. Effects of sire of fetus upon subsequent production and days open of the dam. J Dairy Sci 1977; 60:1964–1969.
56. Servely J-L, Emane MN, Houdebine L-M, Djiane J, Delouis C, Kelly PA. Comparative measurement of the lactogenic activity of ovine placental lactogen in rabbit and ewe mammary gland. Gen Comp Endocrinol 1983; 51:255–262.
57. Thordarson G, Vilalobos R, Colosi P, Southard J, Ogren L, Talamantes F. Lactogenic response of cultured mouse mammary epithelial cells to mouse placental lactogen. J Endocrinol 1986; 109:263–274.
58. Byatt JC, Bremel RD. The lactogenic effect of bovine placental lactogen on pregnant rabbit but not pregnant heifer mammary gland explants. J Dairy Sci 1986; 69:2066–2071.
59. Nagy E, Berczi I, Wren GE, Asa SL, Kovacs K. Immunomodulation by bromocryptine. Immunopharmacology 1983; 102:351–357.
60. Bernton EW, Meltzer MS, Holaday JW. Suppression of macrophage activation and T-lymphocyte function in hypoprolactinemic mice. Science 1988; 239:401–404.
61. Russell DH. Prolactin and immunomodulation. In: Hoshino K, ed. Prolactin gene family and its receptors. Amsterdam: Excerpta Medica, 1988:155–165.
62. Russell DH, Mills KT, Talamantes FJ, Bern HA. Neonatal administration of prolactin antiserum alters the developmental pattern of T-and B-lymphocytes in the thymus and spleen of BALB/c mice. Proc Natl Acad Sci USA 1988; 85:7404–7407.
63. Gibori G, Kalison B, Basuray R, Rao MC, Hunzicker-Dunn M. Endocrine role of the decidual tissue: decidual luteotropin regulation of luteal adenylyl cyclase activity, luteinizing hormone receptors, and steroidogenesis. Endocrinology 1984; 115:1157–1163.
64. Horrobin DF. Prolactin as a regulator of fluid and electrolyte metabolism in mammals. Fed Proc 1980; 39:2567–2570.
65. Colosi P, Marr G, Lopez J, Haro L, Ogren L, Talamantes F. Isolation, purification and characterization of mouse placental lactogen. Proc Natl Acad Sci USA 1982; 79:771–775.

66. Robertson MC, Gillespie B, Friesen HG. Characterization of the two forms of rat placental lactogen (rPL):rPL-I and rPL-II. Endocrinology 1982; 111:1982-1986.
67. Colosi P, Ogren L, Thordarson G, Talamantes F. Purification and partial characterization of two prolactin-like glycoprotein hormone complexes from the midpregnant mouse conceptus. Endocrinology 1987; 120: 2500-2511.
68. Chan JSD, Robertson HA, Friesen HG. The purification and characterization of ovine placental lactogen. Endocrinology 1976; 98:65-76.
69. Becka S, Biletk J, Slaba J, Skarda J, Mikulas I. Some properties of the goat placental lactogen. Experientia 1977; 33:771-772.
70. Hurley TW, Handwerger S, Fellows RE. Isolation and structural characterization of ovine placental lactogen. Biochemistry 1977; 16: 5598-5604.
71. Beckers JF, Fromont-Lienard C, VanDerZwalmen P, Wouters-Ballman P, Ectors F. Isolement d'une hormone placentaire bovine presentant une activite analogue a la prolactine et a l'hormone de croissance. Ann Med Vet 1980; 134:585-595.
72. Murthy GS, Schellenberg C, Friesen HG. Purification and characterization of bovine placental lactogen. Endocrinology 1982; 111: 2117-2124.
73. Eakle KA, Arima Y, Swanson P, Grimek H, Bremel RD. A 32,000-molecular weight protein from the bovine placenta with placental lactogen-like activity in radioreceptor assay. Endocrinology 1981; 110: 1758-1765.
74. Clapp C, Sears PS, Nicoll CS. Binding studies with intact rat prolactin and a 16K fragment of the hormone. Endocrinology 1989; 125: 1054-1059.
75. Kessler MA, Schuler LA. Structure of the bPL gene, and alternative splicing of transcripts. DNA 1991 (in press).
76. Lewis UJ, Bonewald LF, Lewis LJ. The 20,000 dalton variant of human growth hormone: location of amino acid deletions. Biochem Biophys Res Commun 1980; 92:511-516.
77. Lecomte CM, Renard A, Martial JA. A new natural hGH varient—17.5kd—produced by alternative splicing. An additional consensus sequence which might play a role in branchpoint selection. Nucl Acids Res 1987; 15:6331-6348.
78. Hampson RK, Rottman FM. Alternative processing of bovine growth hormone mRNA: nonsplicing of the final intron predicts a high molecu-

lar weight variant of bovine growth hormone. Proc Natl Acad Sci USA 1987; 84:2673-2677.

79. Cooke NE, Ray J, Emery JG, Liebhaber SA. Two distinct species of human growth hormone-variant mRNA in the human placenta predict the expression of novel growth hormone proteins. J Biol Chem 1988; 263:9001-9006.
80. Zieler CG, Kessler MA, Schuler LA. Characterization of a novel prolactin-related protein from bovine fetal placenta. Endocrinology 1990; 126:2377-2382.
81. Shimomura K, Bremel RD. Characterization of bovine placental lactogen as a glycoprotein with N-linked and O-linked carbohydrate side chains. Mol Endocrinol 1988; 2:845-853.
82. Linzer DIH, Lee S-J, Ogren L, Talamantes F, Nathans D. Identification of proliferin mRNA and protein in mouse placenta. Proc Natl Acad Sci USA 1985; 82:4356-4359.
83. Linzer DIH, Nathans D. A new member of the prolactin-growth hormone gene family expressed in mouse placenta. EMBO J 1985; 4:1419-1423.
84. Duckworth ML, Peden LM, Friesen HG. Isolation of a novel prolactin-like cDNA clone from developing rat placenta. J Biol Chem 1986; 261:10879-10884.
85. Duckworth ML, Peden LM, Friesen HG. A third prolactin-like protein expressed by the developing rat placenta: complementary deoxyribonucleic acid sequence and partial structure of the gene. Mol Endocrinol 1988; 2:912-920.
86. Nicoll CS, Mayer GL, Russell SM. Structural features of prolactins and growth hormones that can be related to their biological properties. Endocrinol Rev 1986; 7:169-203.
87. Cunningham BC, Wells JA. High-resolution epitope mapping of hGH-receptor interactions by alanine-scanning mutagenesis. Science 1989; 244:1081-1085.
88. Cunningham BC, Jhurani P, Ng P, Wells JA. Receptor and antibody epitopes in human growth hormone identified by homolog-scanning mutagenesis. Science 1989; 243:1330-1336.
89. Leung DW, Spencer SA, Cachianes G, Hammonds RG, Collins C, Henzel, WJ, Barnard R, Waters MJ, Wood WI. Growth hormone receptor and serum binding protein: purification, cloning and expression. Nature 1987; 330:537-543.
90. Boutin J-M, Jolicoeur C, Okamura H, Gagnon J, Edery M, Shirota M, Banville D, Dusanter-Fourt I, Djiane J, Kelly PA. Cloning and expres-

sion of the rat prolactin receptor, a member of the growth hormone/prolactin receptor gene family. Cell 1988; 53:69–77.

91. Ebbitt DE, Hurley WL, Kessler MA, McDonald DJ, Schuler LA. Characterization of the gene corresponding to bovine placental prolactin-related cDNA I: evolutionary implications. DNA 1989; 8:161–169.

92. Milosavljevic M, Duello TM, Schuler LA. In situ localization of two prolactin-related messenger ribonucleic acids to binucleate cells of bovine placentomes. Endocrinology 1989; 125:883–889.

93. Breathnach R, Chambon P. Organization and expression of eukaryotic split genes coding for proteins. Annu Rev Biochem 1981; 50:349–383.

94. Mitchell PJ, Tjian, R. Transcriptional regulation in mammalian cells by sequence-specific DNA binding proteins. Science 1989; 245:371–378.

95. Murdoch GH, Waterman M, Evans RM, Rosenfeld MG. Molecular mechanisms of phorbol ester, thyrotropin-releasing hormone, and growth hormone. J Biol Chem 1985; 260:11852–11858.

96. Mordacq JC, Linzer DIH. Co-localization of elements required for phorbol ester stimulation and glucocorticoid repression of proliferin gene expression. Genes Dev 1989; 3:760– 769.

97. Owerbach D, Rutter WJ, Martial JA, Baxter JD, Shows TB. Genes for growth hormone, chorionic somatomammotropin, and growth hormone-like gene on chromosome 17 in humans. Science 1980; 209:289–292.

98. Owerbach D, Rutter WJ, Cooke NE, Martial JA, Shows TB. The prolactin gene is located on chromosome 6 in humans. Science 1981; 212:815–816.

99. George DL, Phillips J, Francke U, Seeburg P. The genes for growth hormone and chorionic somatomammotropin are on the long arm of chromosome 17 in region q21 topter. Hum Genet 1981; 57:138–141.

100. Moore DD, Conkling MA, Goodman HM. Human growth hormone: a multigene family. Cell 1982; 29:285–286.

101. Selby MJ, Barta A, Baxter JD, Bell GI, Eberhardt NL. Analysis of a major human chorionic somatomammotropin gene: evidence for two functional promoter elements. J Biol Chem 1984; 259:13131– 13138.

102. Hirt H, Kimelman J, Birnbaum MJ, Chen EY, Seeburg PH, Eberhardt NL, Barta A. The human growth hormone gene locus: structure, evolution and allelic variations. DNA 1987; 6:59–70.

103. Cooke NE, Szpirer C, Levan G. The related genes encoding growth hormone and prolactin have been dispersed to chromosomes 10 and 17 in the rat. Endocrinology 1986; 119:2451–2454.

104. Jackson-Grusby LL,Pravtcheva D, Ruddle FH, Linzer DIH. Chromosomal mapping of the prolactin/growth hormone gene family in the mouse. Endocrinology 1988; 122:2462-2466.
105. Fries R, Beckmann JS, Georges M, Soller M, Womack J. The bovine gene map. Anim Genet 1989; 20:3-29.
106. Womack JE, Moll YD. Gene map of the cow: conservation of linkage with mouse and man. J Hered 1986; 77:2-7.
107. Dietz A, Georges M, Womack JE, Schuler LA. Genes for bovine placental prolactin-related hormones form syntenic group with prolactin (submitted).
108. Hallerman EM, Theilmann JL, Beckmann JS, Soller M, Womack JE. Mapping of bovine prolactin and rhodopsin genes in hybrid somatic cells. Anim Genet 1988; 19:123-131.
109. Fries R, Hediger R, Stranzinger G. Tentative chromosomal localization of the bovine major histocompatibility complex by in situ hybridization. Anim Genet 1986; 17:2987-2994.
110. Skow LC, Womack JE, Petresh JM, Miller WL. Synteny mapping of the genes for 21 steroid hydroxylase, alpha A crystallin, and class I bovine leukocyte antigen in cattle. DNA 1988; 7:143-149.
111. Evans AM, Petersen JW, Sekhon GS, DeMars R. Mapping of prolactin and tumor necrosis factor-beta genes on human chromosome 6p using lymphoblastoid cell deletion mutants. Somat Cell Mol Genet 1989; 15:203-213.
112. Hallerman EM, Nave A, Kashi Y, Holzer Z, Soller M, Beckmann JS. Restriction fragment length polymorphisms in dairy and beef cattle at the growth hormone and prolactin loci. Anim Genet 1987; 18:213-222.
113. Cowan CM, Dentine MR, Ax RL, Schuler LA. Restriction fragment length polymorphisms associated with growth hormone and prolactin genes in Holstein bulls: evidence for a novel growth hormone allele. Anim Genet 1989; 20:157-165.
114. Cowan CM, Dentine MR, Ax RL, Schuler LA. Structural variation around prolactin gene linked to quantitative traits in an elite Holstein sire family. Theor Appl Genet 1990; 79:577-582.
115. Sasavage NL, Nilson JH, Horowitz S, Rottman FM. Nucleotide sequence of bovine prolactin messenger RNA: evidence for sequence polymorphism. J Biol Chem 1982; 257:678-681.
116. Devereux J, Haberli P, Smithies O. A comprehensive set of sequence analysis programs for the VAX. Nucl Acids Res 1984; 12:687-695.

13

Mapping Genes for Resistance to Infectious Diseases in Animals

Harris A. Lewin, Penelope A. Clamp, Jonathan E. Beever, and Lawrence B. Schook

University of Illinois at Urbana–Champaign
Urbana, Illinois

... it is difficult to persuade a farmer to breed for resistance to disease by relating the wonders that have been accomplished with fruit flies and mice. Cows and even chickens command somewhat more respectful attention.

F. B. Hutt, 1958

INTRODUCTION

In human and wild animal populations, natural selection has undoubtedly played a major role in reducing susceptibility to endemic pathogens. The intricate and overlapping mechanisms of immune recognition in higher vertebrates are compelling evidence for this conclusion. In domesticated animals used for food production, breeding practices have been largely directed at increasing product yields with little emphasis on disease resistance. Indeed, there is increasing evidence that general fitness has been decreasing in some breeds of domestic animals. This may be due, in part, to intensive production practices and "genetic stress" imposed by selection (Wu et al., 1989). Animal producers using intensive production systems are becoming increasingly reliant on antibiotics and chemoprophylaxis as methods of disease control. Even though vaccines for many economically important diseases of livestock have been produced, vaccines for a number of important pathogens are not available or economically feasible. With increasing public demands for safe and "natural" food supplies, there is a clear need for improving general resistance to livestock diseases. For practical reasons, it is therefore important to identify genes that influence disease resistance.

One critical limitation in identifying genes that influence resistance to infectious diseases is the underdeveloped genetic maps for all of our farm animal species. For example, cattle have approximately 150 genes mapped to 29 autosomal syntenic groups, of which only 10 have been mapped to specific chromosomes (Womack, this volume). A total of 31 loci have been mapped in sheep and 38 loci have been mapped in pigs (Schook et al., 1990a). These limited maps make identification and mapping of infectious disease loci (IDL) in domestic animals a difficult task at present. However, animal gene mapping research is expanding rapidly, and linkage maps appropriate for the mapping of IDL should soon be available for several species.

Through marker-assisted selection (MAS) and transgenic animal technology coupled with embryo splitting and genotyping, the tools needed to create and propagate disease-resistant livestock are at hand. Furthermore, because of the experimental manipulations that are possible with animals, such as selective breeding and direct challenge with live organisms, domestic animals can serve as useful models for studies of the genetic control of host-parasite interactions in noninbred species. Herein, we review the genetics of resistance to infectious diseases, our experiences with a disease model, and discuss strategies for mapping and utilizing IDL to improve animal health and production.

LOCI INVOLVED IN SUSCEPTIBILITY TO INFECTIOUS DISEASES

Resistance to infectious diseases is an end-point of many genetically controlled physiological processes. Nutritional and environmental stressors, such as temperature and living conditions, interact with host genetic factors to affect disease progression and phenotype. Host resistance to infectious diseases can thus be defined at many levels, each of which is under the control of one or more loci. Disease resistance genes can be broadly classified into three categories (Table 1).

Table 1 Categories of Disease Resistance Genes

1.	Genes for "innate" resistance, such as those that encode receptors for pathogens and genes that control viral replication (e.g., K88, *Fv–1*, and IFNA)
2.	Genes affecting "nonspecific" immunity, such as those regulating phagocytic function and intracellular killing of pathogens (e.g., *Ity*)
3.	Genes affecting "specific" or acquired immunity, such as those influencing antibody and T-cell responses (e.g. IgH, TCR repertoires, and MHC)

Genes Encoding Receptors for Pathogens

In mice, approximately 45 genes have been found to influence the outcome of viral infection (Brinton, 1985). These genes influence all stages of the infective cycle and pathogenesis. Allelic differences at loci affecting pathogen entry and replication have provided important models for the study of host genetic mechanisms involved in resistance to disease. For example, *Rec–1* on chromosome 5 encodes the receptor for murine leukemia virus (Gazdar et al., 1977), and *Fv–1* on mouse chromosome 4 affects the replication of ecotropic Friend leukemia virus (Rowe and Sato, 1973). In swine, the adhesion of enteropathogenic *Escherichia coli* to intestinal brush border is mediated by a locus with two allelic genes *S* and *s*, which has yet to be mapped (Rutter et al., 1975). Piglets with the ss and Ss genotypes are resistant to scours, providing a useful marker of genetic resistance to an economically important disease. In humans, the CD4 gene product acts as a receptor for human immunodeficiency virus (HIV) (Klatzmann et al., 1984). Knowledge of receptors for pathogens can lead to novel strategies for host protection (Sherman et al., 1983).

Loci Affecting "Nonspecific" Immunity

Several genes that affect the microbicidal activity of macrophages and polymorphonuclear leukocytes have been identified (Table 2). The most well characterized is the *Ity* gene, located on mouse chromosome 1 (Plant and Glynn, 1979). Mice homozygous for the susceptibility allele (Ity^s) die from inoculation with ≤10 *Salmonella typhimirium* bacteria,

Table 2 Non-MHC Loci Involved in Susceptibility to Infectious Diseases

Species	Loci	Chromosome	Trait
Mouse[a]	*Cms*	UN	Resistance to *C. immitis*
	Mx	16	Influenza resistance
	xid	X	X-linked immunodeficiency
	Ity	1	Immunity to *Salmonella typhimirium*
	Lsh	1	Immunity to *Leishmania donovani*
	Bcg	1	Immunity to *Mycobacterium bovis, M. lepraemurium, M.intracellulare*
Human[b]	ADA	20	Severe combined immunodeficiency
	CD3	11	Severe combined immunodeficiency
	CD4	12	Severe combined immunodeficiency, AIDS
	CD18	21	Wiskott-Aldrich
	FY	1	Resistance to malaria
	IGH	14	Common variable immunodeficiencies
	IMD1-5	X	X-linked immunodeficiencies
	NP	14	PNP T-cell defect

[a]Summarized from Skamene, 1985.
[b]As reviewed by Hong, 1989.

whereas animals heterozygous or homozygous for the resistance allele (*Ity*[r]) survive challenge with $\geq 10^3$ organisms (O'Brien et al., 19185). Resistant mice appear to be able to restrict growth of *S. typhimirium* within their reticuloendothelial organs. Two linked (or identical) loci, *Lsh* and *Bcg*, regulate the growth of the facultative intracellular pathogens *Leishmania donovani* and *Mycobacterium bovis*, respectively (O'Brien et al., 1980). Recently, genetic resistance to *Brucella abortus* in cattle and *Salmonella abortus ovis* has been shown to be under genetic control and may involve a similar mechanism(s) (Price et al., 1990).

Genes Controlling Specific Immune Responses

Immunologists routinely use quantitative measurements of humoral and cellular immune responses as a measure of immune competence. Levels of responsiveness are generally used as indicators of infectious disease status, depending on the type of response and the antigen in question. For example, the concentration of antibodies to the HIV p24 antigen in serum is correlated with progression of the acquired immunodeficiency syndrome-related complex (Goudsmit et al., 1987). Measures of such immune responses indicate that infection has occurred and often correlate with severity of the disease. Because so many parameters of immune function have been established, mapping of new immune response genes and IDL may not be such a formidable task once adequate linkage maps are at hand.

How many genes control immune responses? The most conclusive breeding experiments aimed at elucidating the number and character of genes affecting immune responsiveness were conducted by Biozzi and his colleagues (1980) with mice. These experiments demonstrated that certain immune responses behave as quantitative traits; i.e., they are determined by the additive effects of several independent loci. For example, the mouse major histocompatibility complex (MHC), or (H-2), accounted for only 12-26% of the interline genetic differences in antibody responses to sheep red blood cells. The number of independent loci estimated to control quantitative levels of antibodies varied from 2 to 11, depending on the selection experiment. The results of this work also indicated that different genes control antibody production and macrophage responses. In general, lines selected for high antibody responses had lower macrophage activity but high cell-mediated immunity (delayed type hypersensitivity and mitogen responsiveness), while selected lines with low antibody activity had high macrophage responses. High responder lines were more resistant to infections that were dependent on antibody immunity (i.e., extracellular organisms) and the low responder lines were more resistant to infections that were dependent on macrophage function (i.e., destruction of intracellular microorganisms).

The Major Histocompatibility Complex

A great deal is known about genes that regulate specific immunological responses, such as antibody and T-cell responses. The most important of these are the genes within the MHC. Many MHC genes have been cloned and sequenced and the three-dimensional structure of a human MHC class I molecule has been determined (Bjorkman et al., 1987a). Polymorphic residues located along the α-helical walls and the β-pleated sheets that form the floor of the antigen binding site account for differences between MHC allomorphs in their ability to bind peptides (Buus et al., 1987). Polymorphic residues located on the outward-facing surfaces of the α-helices are responsible for the phenomenon of MHC restriction and may significantly influence the development of the antigen-specific T-cell repertoire (Bjorkman et al., 1987b).

How, then, can MHC polymorphism influence disease susceptibility? It has been shown experimentally that failure of MHC allomorphs to bind processed antigenic peptide fragments can result in humoral and cellular nonresponsiveness (Buus et al., 1987; Donnermayer and Allen, 1989; Unanue and Cerottini, 1989). Disease susceptibility can result if class I or class II allelic products fail to bind an immunologically relevant peptide, resulting in a breakdown in antigen presentation. Hence, failure to bind immunogenic peptide fragments critical for neutralization of pathogenic agents and immunological clearance of virus-infected cells can result in susceptibility to disease (Bodmer et al., 1989).

Another way in which MHC polymorphism can lead to specific immunological nonresponsiveness to antigens is through the generation of so-called "blind spots" within the T-cell repertoire (Schaeffer et al., 1989). The process of "education" of T-cells to self in the thymus requires specific interaction between T cells and cells bearing self-MHC molecules. When immunogenic segments of peptides mimic self-MHC, or self-MHC plus an endogenous peptide, there may be no response (immunological nonresponsiveness or tolerance) because self-reactive T cells have been eliminated or incapacitated (Marrack et al., 1988). Evidence for the role of blind spots in susceptibility to infectious agents has been mounting (Anderson et al., 1988). Clearly, the gene products of the MHC are the best understood of the IDL in terms of their structure and function, although precisely how MHC alleles influence disease susceptibility in outbred species remains obscure.

MHC-Associated Infectious Diseases in Noninbred Species

In humans there are a few, albeit well-documented immune response (*Ir*) genes. HLA-linked differences in responses to influenza, vaccinia, measles, and several allergens have been demonstrated, and in addition, some studies have been done on HLA associations with infectious diseases (Tiwari and Terasaki, 1985). Data regarding associations between HLA and infectious diseases have been limited mainly because of the obvious difficulties in performing such experiments.

Animal models offer the best possibility to dissect the molecular mechanisms of genetic disease associations. There are several interesting animal models of infectious disease to study, and some are MHC-associated (Table 3). The well-known examples of Marek's disease (Briles et al., 1977) and Rous sarcoma virus-associated tumorigenesis in chickens (Heinzelmann et al., 1981) have provided excellent disease models. MHC effects on scrapie in sheep (Millot et al., 1988), caprine arthritis encephalitis virus infection in goats (Ruff and Lazary,1988), and bovine leukemia virus infection in cattle (Lewin et al., 1988b) have also been described (Table 3). Other infectious diseases for which there is evidence of genetic variation in resistance and susceptibility (Table 3) await further study.

Enzootic Bovine Leukosis

Enzootic bovine leukosis (EBL) is a complex lymphoproliferative disease that has a significant impact on the dairy and beef cattle industries worldwide (Burny et al., 1988). The development of EBL is associated with chronic infection by the bovine leukemia virus (BLV). Some of the

Table 3 Animal Diseases with a Genetic Component to Resistance[a]

Species	Disease	Pathogen	Reference
Cattle	African trypanosomiasis	*Trypanosoma congolense*	Murray *et al.*, 1984
		Trypanosoma vivax	
	Ticks	*Boophilus microplus*	Seifert, 1978
		Ixodid ticks	Francis and Little, 1964
	Enzootic bovine leukosis (EBL)	Bovine leukemia virus	Lewin *et al.*, 1988a,b
	Mastitis	*Staphylococcus* spp.	Emanuelson *et al.*, 1988
		Streptococcus spp.	
		Coliforms	
		Yeast, *Mycoplasma*	
	Brucellosis	*Brucella abortus*	Price *et al.*, 1990
	Nematodes	*Bunostomum phlebotomum*	Esdale et al., 1986
		Cooperia spp.	
		Haemonchus placei	
		Oesophagostomum radiatum	
		Trichostrongylus axei	
Swine	Neonatal Scours	*Escherichia coli*	Rutter *et al.*, 1975
	African Swine Fever	African swine fever virus	Martins *et al.*, 1988

	Atrophic rhinitis	*Bordatella bronchiseptica*	Rothschild *et al.*, 1984
Chicken	Marek's disease	Marek's disease virus	Briles *et al.*, 1977
	Rous sarcoma	Rous sarcoma virus	Heinzelmann *et al.*, 1981
	Avian leukosis	Avian leukosis virus	Crittenden, 1975
	Pullorum	*Salmonella pullorum*	Hutt and Crawford, 1960
	Coccidiosis	*Eimeria tenella*	Lilehoj *et al.*, 1989
	Newcastle's Disease	Newcastle disease virus	Peleg et al., 1976
	Reticuloendotheliosis	Reticuloendotheliosis virus	Scofield et al., 1978
	Infectious bronchitis	Infectious bronchitis virus	Purchase et al., 1966
Horse	Sarcoid	Bovine papilloma virus	Meredith *et al.*, 1986
Sheep and goat	Scrapie	Scrapie	Parry, 1979
	Nematodes	*Trichostrongylus* spp. *Ostertagia* spp. *Haemonchus* spp. *Cooperia* spp.	Outteridge *et al.*, 1988, reviewed in Wakelin, 1978
	Caprine arthritis-encephalitis	Caprine arthritis encephalitis virus	Ruff and Lazary, 1988

[a]The listing of diseases in this table is by no means complete. The literature is full of reports of evidence for genetic resistance to infectious diseases of domestic animals. For an early review see Hutt, 1958.

major features of the pathogenesis of BLV infection include: (1) a variable period between the time of infection and seroconversion to the major viral envelope protein (BLV-gp51); (2) proliferation of virus-infected B cells, which results in a persistent B-cell lymphocytosis (PL) in approximately one-third of BLV- infected cattle; (3) the appearance of antibodies to the viral core protein, p24, which is associated with the development of PL; and (4) clinical lymphosarcoma, which develops in only 1-5% of BLV-infected cattle.

The role of the bovine major histocompatibility complex (BoLA) in EBL has been examined because of the well-documented genetic components of susceptibility to development of PL and tumors. The BLV paradigm provided an ideal opportunity to understand disease resistance mechanisms because of the high prevalence of BLV infection in many dairy herds (often greater than 70% seropositive) and because several subclinical stages of the disease are readily identified (Lewin et al., 1988a). It has now been confirmed in at least three independent herds of Holstein-Friesian cattle that seroconversion to BLV-gp51 and progression to PL are BoLA-dependent. Cows with BoLA-w14 seroconvert at a significantly older age under natural conditions of exposure to the virus than do cows with other BoLA-A alleles. Cows with w14 (and w13) are also "resistant" to development of PL (Fig. 1), whereas cows with BoLA-w12 seroconvert at a younger age and have a significantly higher incidence of PL and greater absolute numbers of B cells in peripheral blood (Lewin et al., 1988b; Lewin, 1989).

Most recent data indicate that resistance and susceptibility to PL maps closely to the BoLA-DRB subregion (van Eijk et al., 1990). BLV-infected, PL- negative cows with the resistant class I phenotypes (w14 and w13) share a 5.9 kb *Taq*I DRB RFLP, whereas PL cows with the susceptibility-associated phenotypes W12 and W15 both have a 3.7-kb *Taq*I DRB RFLP. Results revealed that carriers of the 5.9-kb fragment are resistant to PL ($P < 0.005$), whereas homozygosity for the 3.7-kb fragment is associated with susceptibility ($P < 0.05$), suggesting a dominant mode of inheritance for resistance to PL. These findings support earlier results which implied that the expression of BoLA- DR is affected by BLV infection (Lewin et al., 1987). Thus, resistance and susceptibility to PL may be related to the ability of BLV-infected B cells to present an immunologically relevant epitope to helper T cells.

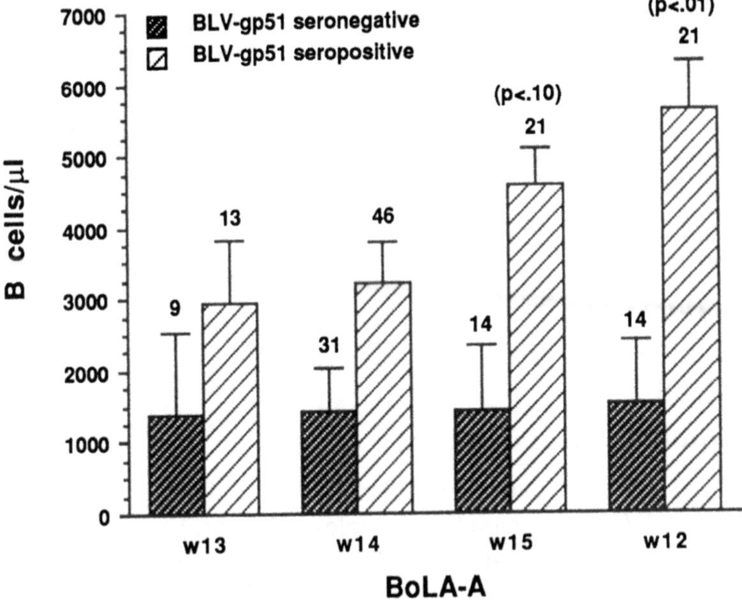

Figure 1 The solid bars show the absolute number of B cells per μl in seronegative cows with the following BoLA-A phenotypes: w13, w14, w15, and w12. There were no differences for any BoLA allele in mean B-cell numbers in seronegative cows. Cows with w12 had significantly greater numbers of B cells than cows with other BoLA-A alleles. Cows with w15 were also susceptible. Although the comparisons were not significant, cows with w13 and w14 tended to have lower numbers of B cells than BLV-infected cows with other BoLA phenotypes. Number above error bars is the number in each category.

The dissection of the EBL paradigm will likely yield important information on the immunological and genetic features of retrovirus-induced and spontaneous neoplasms in noninbred species. Clearly, subclinical progression of BLV infection is under the control of several genes; we have documented that high genetic potential for milk and for fat production is positively associated with the presence of antibodies to BLV-gp51 and the development of PL (Wu et al., 1989). Our studies indicate that the combination of high levels of milk production and BoLA genotype appear to be the strongest influences on susceptibility to B-cell lymphocytosis. Whether there are BoLA- linked genes that influence milk production and/or responses to physiological stressors remains to be determined. Elucidation of such pleiotropic effects of the BoLA system will be important for our overall understanding of the mechanisms of BoLA-linked resistance and susceptibility to infectious diseases.

IDENTIFICATION, MAPPING, AND UTILIZATION OF IDL IN DOMESTIC ANIMALS

Selection of a Disease Model

When choosing an infectious disease model for study, several criteria should be used: (1) the disease has a simple etiology; (2) a sensitive test for the infectious agent is available; (3) the pathogenesis of the disease is well understood, with clearly defined stages of disease progression; and (4) a genetic component to disease susceptibility has been demonstrated. Clinical diseases for which more than one etiological agent is involved, such as mastitis in dairy cattle (Table 3), present difficult theoretical and experimental challenges. Viral diseases appear to be ideal, yet for many animal pathogens diagnosis is often based on insensitive serological tests. Retroviral diseases are particularly problematic in this regard, with infected animals possibly remaining seronegative for extended periods of time. With reliance on a single serological test, it is impossible to distinguish immune individuals from those with advanced infections. The advent of gene amplification using the polymerase chain reaction (PCR) has now made possible the detection of retroviruses in single cells, enabling the accurate diagnosis of infections in which the host is aviremic, such as BLV and HTLV-I. Importantly, if the subclinical stages of infection can be classified and the pathogenesis of the disease is understood, the genetic regulation at each level of the disease process can be determined, from the time of infection until clinical manifestations appear. The ability to identify genetic resistance in noninbred animal models is a distinct advantage over studies with humans.

Experimental Designs

The traditional approaches for identifying IDL in noninbred models are family and population studies. For mapping IDL, linkage analysis has been the predominant method used, employing blood group systems, biochemical polymorphisms, and RFLPs.

Mapping of IDL in domestic animals is accomplished via the production of lines or breeds with resistance and susceptibility to specific diseases. In the past, the cost of producing large farm animals solely for the purpose of studying resistance and susceptibility to specific diseases has been prohibitive because of the long generation intervals and the cost of animal maintenance. However, advanced reproductive biotechnologies, coupled with molecular genetic methodologies, have significantly changed strategies for mapping IDL. For example, embryos and possibly even oocytes can be selected for sex and genotypes using PCR (Lewin et al., 1990). Such genotyped embryos can then be cultured in vitro, microsurgically "split," and transferred to foster mothers. In this way, "cloned" genetically engineered lines of animals homozygous at a particular locus, but differing from other "lines" at one or more alleles, can be produced in a relatively short time. Genotypes can then be evaluated in direct challenge experiments and disease genetics confirmed by classical mating schemes.

The strategy for mapping IDL for specific diseases will depend largely on the character of the disease and how resistance and susceptibility are defined. For example, a program to map the gene(s) responsible for trypanotolerance is presently underway in Africa (Teale, 1990). Bulls selected from the N'Dama breed (*Bos taurus*), known to be tolerant to large parasite burdens (but not resistant to infection), have been crossed with the susceptible Boran breed (*Bos indicus*) to produce full-sib embryo transfer families. After disease phenotypes of the F1 animals are determined, the F2 generation will be produced and also classified for trypanotolerance. The experimental population can then be used for linkage studies. The cross-breeding strategy, utilizing two subspecies, maximizes heterozygosity in the F1, thus increasing the power of the mapping experiment. Certainly, other IDL in domestic animals can be mapped using this strategy. For example, Brahman cattle (*Bos indicus*) are highly resistant to tick infestation (*Boophilus microplus*) compared to *Bos taurus* breeds. Similar strategies are being used to map IDL in Chinese and domestic pig crosses (Schook et al., 1990b).

Other genetic manipulations can also be used to identify and characterize IDL. For example tetra- and hexaparental chimeras can be used to evaluate the role of tolerance induction in creating potential blindspots in the T-cell repertoire for immunodominant antigens on pathogenic organisms. Birds represent another unique biological material that can be exploited for identification and characterization of IDL. In chickens, aneuploidy for the chicken microchromosome carrying the MHC has been reported (Bloom and Bacon, 1985). Aneuploidy and polyploidy may be an interesting means of evaluating gene dosage effects on disease resistance.

Comparative Mapping Strategy

It is abundantly clear that conservation of linkage is a fundamental phenomenon among higher vertebrates (Womack, this volume). Homologous IDL can be identified in different species through the comparative mapping approach (Fig. 2). Using the heterologous gene or a closely linked flanking marker, the IDL can be mapped by linkage studies and in situ hybridization. Linkage studies using flanking markers would require disease challenge or an in vitro assay that reflects IDL genotype (Price et al., 1990).

Utilization of Mapped IDL

A desirable outcome of IDL mapping would be the production of livestock with general resistance to infectious diseases. In practice, this may be very difficult to achieve because of the wide variety of organisms, environments, and physiological factors involved in this trait. It appears that breeding for genetic resistance to specific diseases is warranted under certain conditions. For example, introduction of the trypanotolerance gene into the more productive European and native breeds would be of great economical consequence to developing countries in western Africa. Other types of diseases where selection would be applicable would be those like EBL, for which an effective or economical vaccine is difficult or expensive to produce. MHC-linked resistance to infectious agents is usually a dominant trait because MHC alleles are codominantly expressed. The utilization of sires with resistant

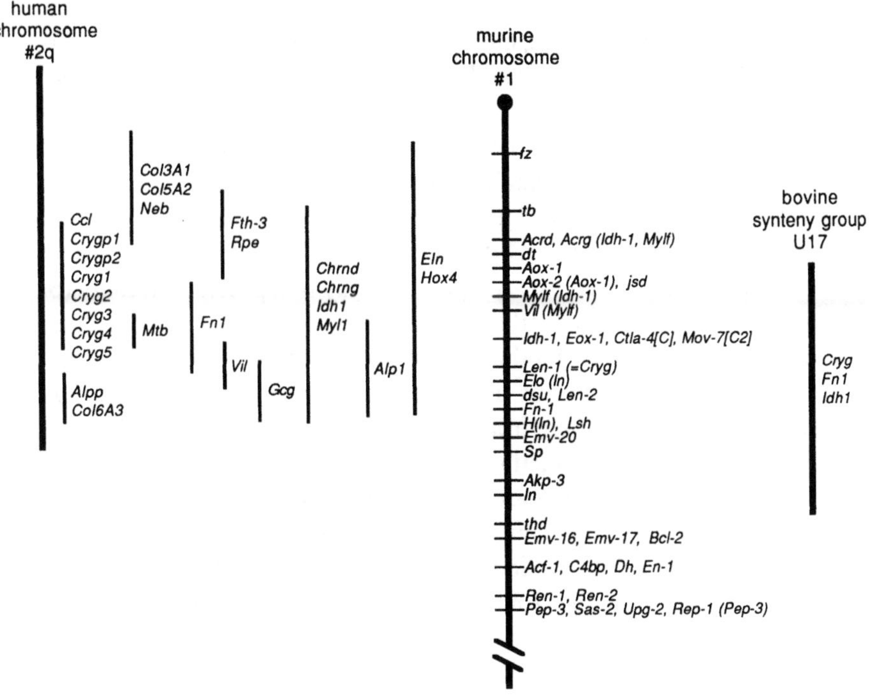

Figure 2 Conserved syntenic groups of human, mouse, and cattle. These maps demonstrate how conserved syntenic groups can be used to identify genes associated with disease susceptibility. The *Lsh* locus, which controls resistance to *Leishmania donovani* infection in mice, belongs to a syntenic group that is conserved in three species.

haplotypes (i.e., with BoLA w13 and w14 for BLV) would therefore be a sensible strategy for reducing potential economic losses due to subclinical BLV infection; particularly in herds with a high incidence of infections. This strategy would permit cows that are genetically resistant to the development of PL to also fully express their genetic potential for milk and milk fat production.

The development of appropriate herd health data bases will make it possible to include immunological functions into classical selection indices. Improvement for specific resistance or specific immune functions can be achieved by marker-assisted selection (Weller and Fernando, this volume). In pigs, multitrait selection experiments are underway (B. Mallard, B. Wilkie, and B. Kennedy, personal communication) to examine whether improved immune function can be selected for without negative effects on the classical quantitative production traits.

Finally, direct genetic manipulations are also possible strategies to increase disease resistance. Transgenic animals and chimeras are being produced for this purpose (Crittenden, 1986). Transgenics expressing antisense mRNA for critical viral proteins may be a possible strategy for producing resistance to specific viral diseases (Zamencnic and Stephenson, 1978). In the future, modulation of IDL in vitro (e.g., cytokine loci) may provide a direct means of influencing resistance to specific diseases, even in different environments.

With advanced reproductive biotechnologies, it is now possible to spread desirable IDL alleles and transgenes by animal cloning, although we would be wise to first consider the consequences of limiting genetic diversity. As more IDL are identified and mapped, and their functions understood, new strategies for increasing disease resistance will emerge that are unthinkable even today.

REFERENCES

Anderson DC, van Schooten WCA, Barry ME, Janson AAM, Buchanan TM, de Vries RRP. A *microbacterium leprae*-specific human T cell epitope cross reactive with an HLA-DR2 peptide, Science 1988; 242:259–261.

Biozzi G, Siquera M, Stiffle C, Ibanez OM, Mouton D, Ferreira VCA. Genetic selections for relevant immunological functions. In: Fougereau M., Dausset J, eds. Immunology 80. Progress in Immunology IV. New York: Academic Press, 1980:432–457.

Bjorkman PJ, Saper MA, Samraoui B, Bennett WS, Strominger JL, Wiley DC. Structure of the human class I histocompatibility antigen, HLA-A2. Nature 1987a; 329:506–511.

Bjorkman PJ, Saper MA, Samraoui B, Bennett WS, Strominger JL, Wiley DC. The foreign antigen binding site and T cell recognition regions of class I histocompatibility antigens. Nature 1987b; 329:512–518.

Bloom SE, Bacon LD. Linkage of the major histocompatibility (B) complex and the nucleolar organizer in the chicken. J Hered 1985; 76:146–154.

Bodmer HC, Gotch FM, McMichael AJ. Class I cross-restricted T cells reveal low responder allele due to processing of viral antigen. Nature 1989; 337:653-655.

Briles WE, Stone HA, Cole RK. Marek's disease: effects of B histocompatibility alloalleles in resistant and susceptible chicken lines. Science 1977; 195:193- 195.

Brinton MA. Genetic control of resistance of viral infections. In: Skamene E, ed. Genetic control of host resistance to infection and malignancy. New York: Alan R Liss, 1985:111-123.

Burny A, Kettmann R, Mammerickx M, Portetelle D, Willems L, Cleuter Y, Marbaix C, van den Broeke A, Thomas R. Bovine leukemia: facts and hypotheses derived from the study of an infectious cancer. Vet Microbiol 1988; 17:197- 218.

Buus S, Sette A, Colon SM, Grey HM. The relation between major histocompatibility complex (MHC) restriction and the capacity of Ia to bind immunogenic peptides. Science 1987; 235:1353-1358.

Crittenden LB. Two levels of genetic resistance to lymphoid leukosis. Avian Dis 1975; 19:281-292.

Crittenden LB. Identification and cloning of genes for insertion. Poultry Sci 1986; 65:1468-1473.

Donnermayer DL, Allen PM. Binding to Ia protects peptides from proteolytic degradation. J Immunol 1989; 142:1063-1068.

Emmanuelson U, Dannel B, Philipsson J. Genetic parameters for clinical mastitis, somatic cell counts, and milk production estimated by multiple-trait restricted maximum likelihood. J Dairy Sci 1988; 71:467-476.

Esdale CR, Leutton RD, O'Rourke PK, Rudder TH. The effect of sire selection for helminth egg counts on progeny helminth egg counts and live weight. Proc Aust Soc Anim Prod 1986; 16:199-202.

Francis J, Little DA. Resistance of droughtmaster cattle to tick infestation and babesiosis. Aust Vet J 1964; 40:247-253.

Gazdar AF, Oie H, Lalley P, Moss WW, Minna JD. Identification of mouse chromosomes required for murine leukemia virus replication. Cell 1977; 11:949- 956.

Goudsmsit J, Lange JMA, Paul DA, Dawson G. Antigenemia and antibody titers to core and envelope antigens in AIDS, AIDS-related complex, and subclinical immunodeficiency infection. J Infect Dis 1987; 155:558-560.

Heinzelmann EW, Zsigray RM, Collins WM. Cross reactivity between RSV-induced tumor antigen and B5 MHC alloantigen in the chicken. Immunogenetics 1981; 13:29-37.

Hong R. Immunodeficiency diseases. In: Litwin SD, ed. Human immunogenetics: basic principles and clinical relevance. New York: Marcel Dekker, 1989:257- 284.

Hutt FB. Genetic resistance to disease in domestic animals. Ithaca, NY: Comstock Publishing Associates, Cornell University Press, 1958:198.

Hutt FB, Crawford RD. On breeding chicks resistant to pullorum disease without exposure thereto. Can J Genet Cytol 1960; 2:357-370.

Klatzmann D, Champagne E, Chamare S, Gruest J, Guetard D, Hercend T, Gluckman J-C, Montagnier L. T-lymphocyte T4 molecule behaves as the receptor for human retrovirus LAV. Nature 1984; 312:767-768.

Lewin HA. Disease resistance and immune response genes in cattle: strategies for their detection and evidence of their existence. J Dairy Sci 1989; 72:1334-1348.

Lewin HA, Nolan TJ, Schook LB. Altered expression of class II antigens on peripheral blood B lymphocytes from BLV-infected cows with persistent lymphocytosis. In: Schook LB, Tew JG, eds. Antigen presenting cells: diversity, differentiation, and regulation. New York: Alan R Liss 1987: 211- 220.

Lewin HA, Wheeler MB, Schook LB. In: UFO-PCR: A strategy for linkage mapping in species with underdeveloped genetic maps. In: Womack JE, ed. Mapping the genomes of agriculturally important animals. New York: Cold Spring Harbor Laboratory Press, 1990:95-102.

Lewin HA, Wu MC, Stewart JA, Nolan TJ. Peripheral B lymphocyte percentage as an indicator of subclinical progression of bovine leukemia virus infection. J Dairy Sci 1988a; 71:2526-2534.

Lewin HA, Wu MC, Nolan TJ, Stewart JA. Association between BoLA and subclinical bovine leukemia virus infection in a herd of Holstein-Friesian cows. Immunogenetics 1988b; 27:338-344.

Lillehoj HS, Ruff MD, Bacon LD, Lamont SJ, Jeffers TK. Genetic control of immunity to *Eimeria tenella*. Interaction of MHC genes and non-MHC linked genes influences levels of disease susceptibility in chickens. Vet immunol Immunopathol 1989; 20:135-148.

Marrack P, Lo D, Brinster R, Palmiter R, Burkly L, Flavell RH, Kappler J. The effect of thymus environment on T cell development and tolerance. Cell 1988; 53:627-634.

Martins C, Mebus C, Scholl T, Lawman M, Lunney J. Virus-specific CTL in SLA- inbred swine recovered from experimental African swine fever virus (ASFV) infection. In Cytotoxic T cells. Ann NY Acad Sci 1988; 532:462-464.

Meredith D, Elser AH, Wolf B, Soma LR, Donawick WJ, Lazary S. Equine leukocyte antigens: relationships with sarcoid tumors and laminitis in two pure breeds. Immunogenetics 1986; 23:221-225.

Millot P, Chatelin J, Dautheville C, Salmon D, Cathala F. Sheep major histocompatibility (OLA) complex: linkage between scrapie susceptibility/resistance and the OLA complex in Ile-de-France sheep progenies. Immunogenetics 1988; 27:1-11.

Murray M, Trail JCM, Davis CE, Black SJ. Genetic resistance to African trypanosomiasis. J Infect Dis 1984; 149:311-319.

O'Brien AD, Rosenstreich DL, Taylor BA. Control of natural resistance to *Salmonella typhimurium* and *Leishmania donovani* in mice closely linked by distinct genetic loci. Nature 1980; 287:440-442.

O'Brien AD, Weinstein DL, Lissner CR, Winstanley FP, Huppi K. *Ity* gene expression in resident peritoneal macrophages isolated from *Salmonella typhimurium*-resistant and *Salmonella typhimurium*-susceptible congenic mice. In: Skamene E. ed. Genetic control of host resistance to infection and malignancy. New York: Alan R. Liss 1985:229-237.

Outteridge PM, Windon RG, Dineen JL. An ovine lymphocyte antigen marker for acquired resistance to the *Trichostrongylus colubriformis*. Int J Parasitol 1988; 18:853-858.

Parry HB. Elimination of natural scrapie in sheep by sire genotype selection. Nature 1979; 277:127-129.

Peleg BA, Soller M, Ron N, Hornstein K, Brody T, Kalmar E. Familial differences in antibody response of broiler chickens to vaccination with attenuated and inactivated Newcastle disease virus vaccine. Avian Dis 1976; 20: 661-668.

Plant J, Glynn AA. Locating *Salmonella* resistance gene on mouse chromosome 1. Clin Exp Immunol 1979; 37:1-6.

Price RE, Templeton JW, Smith R III, Adams LG. Ability of mononuclear phagocytes from cattle naturally resistant or susceptible to brucellosis to control in vitro intracellular survival of *Brucella abortus*. Infect Immun 1990; 58:879-886.

Purchase HG, Cunningham CH, Burmester BR. Genetic differences among chicken embryos in response to inoculation with an isolate of infectious bronchitis virus. Avian Dis 1966; 10:162-172.

Rothschild MF, Chen HL Christian LL, Lie WR, Venier L, Cooper M, Briggs C, Warner CM. Breed and swine lymphocyte antigen haplotype differences in agglutination titers following vaccination with *Bordatella bronchiseptica*. J Anim Sci 1984; 59:643-649.

Rowe WP, Sato H. Genetic mapping of the *Fv–1* locus of the mouse. Science 1973; 180:640-641.

Ruff G, Lazary S. Evidence for linkage between the caprine antigen (CLA) system and susceptibility to CAE virus-induced arthritis in goats. Immunogenetics 1988; 28:303-309.

Rutter JM, Burrows MR, Sellwood R, Gibbons RA. A genetic basis for resistance to enteric disease caused by *E. coli.* Nature 1975; 257:135-136.

Schaeffer EB, Sette A, Johnson DL, Bekoff MC, Smith JA, Grey HM, Buus S. Relative contribution of "determinant selection" and "holes in the T-cell repertoire" to T-cell responses. Proc Natl Acad Sci USA 1989; 86: 4649-4653.

Schook LB, Beever JE, Clamp PA, Lewin HA, McLaren DG. Status of the pig gene map. In: Womack JE, ed. Mapping the genomes of agriculturally important animals. New York: Cold Spring Harbor Laboratory Press, 1990a: 123-130.

Schook LB, McLaren DG, Lewin HA. Molecular biology and classical breeding: integration to accelerate the rate of genetic change in swine. Anim Biotechnol 1990a (in press).

Scoffield VL, Spencer JL, Briles WE, Bose HR. Differential mortality and lesion response to reticuloendotheliosis virus infection in Marek's disease-resistant and susceptible chicken lines. Immunogenetics 1978; 7:169-172.

Seifert GW. Variations between and within breeds of cattle in resistance to field infestation of the cattle tick (*Boophilus microplus*). Aust J Agric Res 1978; 22:159-168.

Sherman DM, Acres SD, Sadowski PL, Springer JA, Bray B, Raybould TJG, Muscoplat CC. Protection of calves against fatal enteric colibacillosis by orally administered *Escherichia coli* K99-specific monoclonal antibody. Infect Immun 1983; 42:653-658.

Skamene E, ed. Genetic control of host resistance to infection and malignancy. New York: Alan R Liss 1985:848 pp.

Teale AJ. Genes controlling disease resistance as targets of bovine genome research, with particular reference to trypanotolerance. In: Womack JE, ed. Mapping the genomes of agriculturally important animals. New York: Cold Spring Harbor Laboratory Press, 1990:65-72.

Templeton JW, Smith R. III, Adams LG. Natural disease resistance in domestic animals. J Am Vet Med Assoc 1988; 192:1306-1314.

Tiwari JL, Terasaki PI. HLA and disease associations. New York: Springer-Verlag, 1985:472.

Unanue E, Cerottini J-C. Antigen presentation. FASEB J 1989; 3:2496-2502.

Unanue E, Cerottini J-C. Antigen presentation. FASEB J 1989; 3:2496-2502.

van Eijk MJT, Xu A, Muggli-Cockett NE, Lewin HA. Association between BoLA-DRB genotypes and development of persistent lymphocytosis in bovine leukaemia virus-infected cows (abstract). Anim Genet 1990 (in press).

Wakelin D. Genetic control of susceptibility and resistance to parasitic infections. Adv Parasitol 1978; 16:219-308.

Wu MC, Shanks RD, Lewin HA. Milk and fat production in dairy cattle influenced by advanced subclinical progression of bovine leukemia virus infection. Proc Natl Acad Sci USA 1989; 86: 993-996.

Zamencnic PC, Stephenson ML. Inhibition of Rous sarcoma virus replication and cell transformation by a specific oligodeoxynucleotide. Proc Natl Acad Sci USA 1978; 75:280-284.

14

Strategies for the Improvement of Animal Production Using Marker-Assisted Selection

Joel I. Weller

The Volcani Center
Bet Dagan, Israel

Rohan L. Fernando

University of Illinois at Urbana–Champaign
Urbana, Illinois

INTRODUCTION

During the last 50 years genetic improvement has dramatically increased the economic efficiency of many domestic animals. Nearly all of the gain can be attributed to application of principles of quantitative genetics and statistics. Understanding of the physiology of mendelian genetics of traits of economic importance has been nearly irrelevant to the methods employed to achieve genetic improvement in most livestock species. Genetic markers can be used to identify and genetically manipulate the individual quantitative trait loci (QTL) that affect quantitative traits. The objective of this chapter is to describe techniques to utilize genetic markers in commercial animal breeding, and to estimate the expected gains from marker-assisted selection (MAS). We will first describe briefly the current methodologies and achievements of practical animal breeding. We will then discuss the limitations of traditional methods and explain how MAS may be useful in situations where traditional methods have not been. We will then describe the steps necessary to employ MAS on commercial animal populations with emphasis on location of QTL affecting economic traits, and elaborate on advantages of MAS for within- and between-breed genetic improvement.

TRADITIONAL METHODS OF SELECTION FOR QUANTITATIVE TRAITS

Review of Traditional Selection Methodology

In animal breeding, selection procedures have been based on an additive genetic model where the genotype value is the sum of the effects of the individual genes. The joint effects of genes within a locus (dominance) and across loci (epistasis) are usually ignored in selection theory and practice.

It can be shown that, under additive genetic inheritance, selecting animals with the highest genotypic values will lead to maximum genetic progress. However, for most traits of economic importance, the genotypic value of an individual cannot be observed, and selection decisions have to be based on phenotypic data. Thus, ideally, animals are ranked according to the regression of their genotypic values on phenotypic data, and the highest-ranking animals are selected. When a constant number of animals is to be selected, this procedure has been shown to maximize expected genotypic progress (1-3).

Under additive genetic inheritance, the genotypic value is approximately normally distributed because it is the sum of a large number of gene effects, each being small (4). The environmental deviation is also due to a large number of factors, and thus, the phenotype is also approximately normally distributed. When genotypic and phenotypic values are normally distributed, the regression of genotypic values on phenotypic values is linear.

Let u be a vector of unobservable genotypic values, and let y be a vector of phenotypic data. Then, the linear regression of u on y, $\hat{u}$, is

$$\hat{u} = E(u) + CV^{-1}[y - E(y)] \quad (1)$$

where E(.) stands for expected value or mean, C is the covariance matrix between u and y, and V^{-1} is the inverse of the covariance matrix of y. From Eq. (1), it can be seen that calculation of û requires E(u), E(y), C, and V to be known. Henderson (5–7) has shown that substituting the generalized least-squares estimate (GLS) of E(u) and E(y) for their true values in Eq. (1) gives the best linear unbiased predictor (BLUP) of u. Goffinet (1) and Fernando and Gianola (3) have discussed the optimality of this approach for selection. In practice, estimates of C and V also have to be substituted for their true values in Eq. (1). Estimates of C and V can be obtained from phenotypic data or relatives.

The classical methods for estimation of these variance covariance parameters from field data were developed by Henderson (8). Since then, other methods with superior properties have become available (7,9–12). Gianola et al. (13) have discussed the prediction of u when C and V are unknown.

The order of V in livestock applications can be very large, ranging from a few thousands to millions (14,15). Thus, use of Eq. (1) is not computationally feasible, and BLUP of u is usually calculated using Henderson's mixed model equations (MME) (6).

Suppose, for example, y can be modeled as

$$y = X\beta + Zu + e \quad (2)$$

with $E(y) = X\beta$, $E(u) = 0$, $E(e) = 0$, $Var(u) = A\sigma_u^2$, $Var(e) = I\sigma_e^2$, and $Var(y) = (ZAZ'\sigma_u^2 + I\sigma_e^2)$, where Var(.) stands for variance. Then, the MME for obtaining BLUP are

$$\begin{bmatrix} X'X & X'Z \\ Z'X & Z'Z + A^{-1}\sigma_e^2/\sigma_u^2 \end{bmatrix} \begin{bmatrix} \tilde{\beta} \\ \tilde{u} \end{bmatrix} = \begin{bmatrix} X'y \\ Z'y \end{bmatrix} \quad (3)$$

Table 1 Estimates of Genetic Trends for Some Economic Traits Under Selection in Domestic Animals

Species	Trait	Genetic trend/year	Ref.
Dairy cattle	Milk production	84 kg	18
		102 kg	14
	Fat production	2.3 kg	14
Beef cattle	Yearling weight	0–2 kg	19
Sheep	Carcass weight	0.24 kg	20
Goats	Milk production	2 kg	21
Swine	Backfat	-0.19 mm	22
	Days to 90 kg	-1.72 d	22

where $\tilde{\beta}$ is GLS of β and $\tilde{u}$ is BLUP of u. The inverse of A in Eq. (3) is very efficiently obtained using an algorithm described by Henderson (16). The MME are usually solved by iterative procedures such as Gauss–Siedel iteration. Recently iterative algorithms have been described to solve the MME without setting them up directly (17), and systems with over 20 million equations have been solved successfully (15).

Advantages and Disadvantages of Traditional Animal Breeding

Some estimates for recent genetic trends for traits of economic importance in several species are presented in Table 1.

Genetic trends for milk production in developed countries approach 100kg/year, or slightly more than 1%/year. Genetic trends have been similarly impressive for sheep, goats, and swine, but not for beef cattle, which has consistently lagged behind dairy cattle with respect to genetic improvement. This is one case where traditional methods of breeding have not been successful. The limitations of these methods will now be reviewed.

Many traits of economic importance have low heritability and therefore have not been emphasized in selection programs. Prime examples are most traits related to fertility and disease resistance. Traditional selection has worked best on traits with normal or quasi-normal distributions. Thus traditional breeding programs have emphasized production traits at the expense of non-production traits, which are very often categorical. Again, fertility and disease-related traits are prime examples. Also included are traits with continuous, but highly asymmetric distributions, such as milk somatic cell count. Trait-based selection is less effective when the trait of interest is expressed in only one sex, such as milk production, or number of offspring per litter. The situation is most extreme in dairy cattle, in which the males have nearly unlimited fertility, but do not express the production traits of interest, while female fertility, and therefore possibilities for selection, are extremely limited. (New developments in biotechnology, such as relatively inexpensive techniques for embryo transfer and splitting, may radically increase the potential number of offspring per cow.) Finally, selection has generally been limited to those traits which can be conveniently measured on large samples of animals. Many traits of economic importance, such as individual feed consumption, are prohibitively expensive to measure under conditions of commercial production. Some traits, such as survival rate, are expressed too late in the animal's life to be useful criteria for selection, and other traits, such as body composition, can be measured only after slaughter.

Traditional selection techniques have not been able to increase the economic efficiency of beef cattle, because the traits with definite economic value, such as feed consumption or fertility traits, are either difficult to measure or have low heritability. In addition, paternity of most animals is generally unknown and thus precludes the ability to select along the sire path. However, even when paternity is known, evaluation can only be performed for those traits that are measured, which generally consist of weights at certain ages, and it is not at all clear that selection for increased growth rate is desirable.

As shown above, traditional selection is effective only on additive genetic variation. Thus dominance and epistatic variation are generally not utilized to increase economic efficiency. Traditional selection within populations has also not been very effective when the breeding goal has been to improve several traits with negative genetic correlations among them. A prime example is milk production and fat percent.

Although significant progress has been made for milk production, which has been the primary objective, the genetic trend for fat percent has been either close to zero or negative, due to a high negative genetic correlation between these traits, despite that fact that fat percent has high heritability and a quasinormal distribution (14).

Traditional methods are also inefficient in crossbreeding. The three main goals of crossbreeding are: (1) heterosis, (2) increased genetic variation, and (3) introgression. "Heterosis" in agricultural species is generally defined as a situation in which the progeny of a cross between two breeds is economically more favorable than either parental strain. As shown by Moav (23), heterosis consists of at least five different biological phenomena. In general, heterosis can consist either of a superior value for the offspring for a specific trait or a combination of different desirable traits from the parental lines; for example high growth rate from a male line and high fertility from a female line. Heterosis for a specific trait is usually ascribed to overdominance at specific loci. Generally, the most favorable breed combination has been found by trial-and-error among a large number of possible combinations. Furthermore, once the most desirable combination is determined, traditional methods of selection do not provide avenues for further improvement.

Upon occasion, different breeds are crossed to produce a population with increased genetic variance for traits of interest. Once the cross is produced, traditional selection methods can be used to increase the trait value in the desired direction. However, since traditional methods consider only trait values, desirable genes may be lost through selection. Generally only the "best" breeds will be considered as parental candidates. Thus a particular breed that is inferior for the major economic traits will not be considered, even though this breed may still have superior alleles for some genes that affect the economic traits, and this "cryptic" genetic variation is not utilized.

Often the breeder would like to transfer an allele for a specific trait from strain A to strain B. This can be done by a process called "introgression." This entails first crossing the two strains, followed by a series of backcrosses to strain B. At each generation, only those progeny which have retained the desirable allele are used as parents for the next generation. After several generations, a progeny group can be produced which have retained the desired allele on a genetic background nearly identical to that of strain B. Finally, progeny of the last backcross generation are mated among themselves to produce individuals homozygous for the desired allele. In addition to being time-consuming and tedious, introgression can generally be done only if the genotypes for the specific loci in question can be identified in the progeny. Thus the Booroola gene for fecundity in sheep can be introgressed into other breeds, because it is possible to determine progeny genotype for this locus (24). However, other high-fertility sheep are not useful for this purpose, because the loci that determine this trait are unknown and therefore cannot be identified in the backcross progeny.

IMPLEMENTATION OF MARKER-ASSISTED SELECTION

Generally two steps are required for implementation of MAS: location of loci affecting the traits in question and manipulation of these QTL with the aid of genetic markers. We will first review the experimental designs and statistical methods that have been used or suggested for use to locate QTL. We will then describe the possible scenarios in which manipulation of QTL may be more effective than traditional methods and present estimates of the expected gain, where available. Scenarios for within- and between-breed QTL manipulation will be discussed separately.

Experimental Design for Location of Quantitative Trait Loci

The main difficulty in location of QTL is that their effect will generally be small with respect to the other sources of variation in the population. Without the aid of genetic markers, genotype identification of QTL is possible only for loci with effects greater than a phenotypic standard deviation (25,26). However, for a trait with heritability of 0.2, it is unlikely that any single QTL will be of this magnitude. For example, a QTL with two alleles of equal frequency, and codominance that accounts for 1% of the phenotypic variance (5% of the additive genetic variance), will have a mean difference between the homozygotes of 0.28 phenotypic standard deviations (27). Loci of this magnitude cannot be discerned by inspection of the phenotypic distribution. The statistical

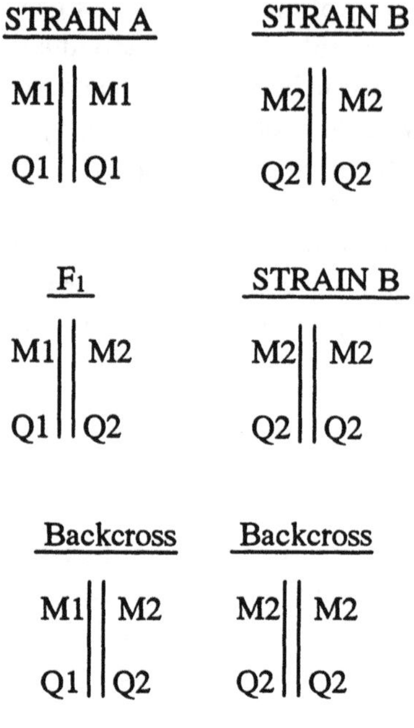

STRAIN A — M1 Q1 / M1 Q1

STRAIN B — M2 Q2 / M2 Q2

F_1 — M1 Q1 / M2 Q2

STRAIN B — M2 Q2 / M2 Q2

Backcross — M1 Q1 / M2 Q2

Backcross — M2 Q2 / M2 Q2

Figure 1 A backcross design for two inbred lines differing in both a marker locus and a linked QTL.

distribution of the trait values of a sample of 1000 individuals will not be significantly different from normality.

Genetic markers can be used to locate QTL of this size if there is relatively tight genetic linkage between the marker and the QTL and linkage disequilibrium between the two loci. This is illustrated in Figure 1 for the case of a cross between two inbred lines followed by a backcross to one of the parental strains. The marker locus M is linked to the QTL denoted Q. Strain A is homozygous for allele M1 of the marker locus and allele Q1 of the QTL, while strain B is homozygous for allele M2 of the marker locus and allele Q2 of the QTL. Only the genotype for the marker locus can be determined by inspection of individuals. The two parental strains may be different or similar for the quantitative trait in question, and this difference may be due to many other loci in addition to Q. These strains are crossed to produce the F_1, which will be heterozygous for the complete genome, including the marker locus and the linked QTL. In this design the F_1 is backcrossed to strain B. Two genotypes will be produced with respect to locus M, M1M2 and M2M2. When the mean values of these two groups of individuals are compared, all sources of genetic variation will be distributed randomly, except for the chromosomal segment linked to the locus M. Thus a significant difference between the backcross genotypes indicates a segregating QTL linked to M.

Similar results can be derived from production of an F_2 from two inbred lines, except in this case, three genotypes will be produced, and it is possible to estimate the dominance effect of the QTL. The effect measured will be proportional to 1-r in the backcross, and 1-2r in the F_2, where r is the frequency of recombination between the two loci. The power of these experiments is increased dramatically if pairs of linked markers (marker brackets) are used (28–30). The optimal situation for these experiments is crosses between lines with genetic markers spaced at intervals of about 20 map units. Then any segregating QTL should be tightly linked to at least one genetic marker (31). This has been denoted "saturation of the genetic map with marker loci" (32). Although animal strains with saturated maps are not currently available, restriction fragment length polymorphisms (RFLP) have made this a reality for certain plant species (33–38). RFLPs have also been found in domestic animal populations (39–42).

Experiments of this type have been performed with plants and laboratory animals (30,43–48). In domestic animals inbred lines are generally not available. Furthermore, even in the few cases where inbred lines have been produced, it is generally prohibitively expensive to create the F_2 or backcross populations described above. Most attempts to locate QTL in domestic animals have concentrated on analysis of existing populations. Although it is expected that there will generally be linkage equilibrium between genetic markers and linked QTL, a number of situations can be postulated in which this is not the case: for example, if the genetic marker is the result of a recent mutation that has not yet reached equilibrium in the population,or if the QTL is a pleitropic effect of the genetic marker. It is likely that in many of the early experiments, which used morphological markers to detect QTL, the significant effects found may have been pleiotropic effects of the marker loci (46–48).

If there is linkage disequilibrium between marker loci and QTL, then the QTL effect can be detected by the relatively simple approach of estimating the effect of various genetic markers on different quantitative traits (49–51). Since linkage disequilibrium is likely to be rare, it is not surprising that this approach has generally not been successful. Another problem with many analyses of field data has been that the effects of a number of genetic markers on several traits were analyzed as separate experiments, without accounting for this situation in determination of significance levels. For example, if the effect of 10 markers is tested on 10 quantitative traits, then the total number of marker–trait combinations is 100. If all these combinations are then analyzed as separate experiments, the expectation is that five of these combinations will show significant effects at the 5% level purely by chance (52). Although methods have been developed to account for this problem, they have generally not as yet been applied in practice (53). Thus many of the effects that were reported in the literature have not been repeatable.

Since large half-sib families are available for a number of domestic animals, several studies have suggested analyses of the progeny of heterozygous sires (54–61). Since there will be linkage disequilibrium between the QTL and the genetic marker within the progeny of each sire, but not among sires, the following analysis model can be used to detect the presence of QTLs segregating within half-sib families:

$$y_{ijk} = S_i + M_j + SM_{ij} + e_{ijk} \tag{4}$$

where y_{ijk} is the quantitative trait value for the kth progeny of the ith sire with the jth genotype for the genetic marker, S_i is the effect of the ith sire, M_j is the effect of the jth marker genotype, SM_{ij} is the interaction effect between the ith sire and the jth marker genotype, and e_{ijk} is the random residual effect. A significant sire by marker effect is indicative of a segregative QTL linked to the genetic marker. Since the main marker effect is itself not important, assuming linkage equilibrium, an alternative model has been suggested with marker effects nested within sires.

An alternative to the "daughter" design described above is the "granddaughter" design, in which genetic marker genotype is determined on sons of a heterozygous sire, while trait values are scored on the granddaughters (52). To achieve equal power to the daughter design, it is necessary by the granddaughter design to score many more individuals for quantitative traits, but far fewer for marker genotype. In many cases this scheme may be preferable, since scoring for quantitative traits may be much less expensive than determining RFLP genotype. Furthermore, for many populations of domestic animals, existing data banks on quantitative traits can be used, so that the cost of obtaining data for these traits is virtually nil.

Statistical Methodologies for Detection of Quantitative Trait Loci

The following assumptions have generally been employed in the analysis of QTL-marker experiments:

1. No more than one segregating QTL is linked to the genetic marker.
2. The underlying trait distribution is normal.
3. The QTL affects the trait mean, but not higher-order moment, such as, the variance or skewness.

The first assumption has not been seriously challenged, other than the heuristic argument that if major QTL are relatively rare, then the probability of two linked QTL will be low (30). Deviation from normality can easily be tested (62). If the distribution is continuous, but does deviate significantly from normality, it is often possible by algebraic transformation to obtain a quasinormal distribution (63). Although most studies have assumed that QTL genotype does not affect the trait variance, several studies have found significant QTL variance effects (30,45,47,48). Methods that do not require the assumption of no QTL variance effect have been developed (47,64).

Most studies have used simple linear models and analysis of variance (ANOVA) to test for significance of QTL effects. ANOVA has the advantages of being readily available in many statistical packages and has well-defined properties. However, several studies have shown that for incomplete linkage, ANOVA does not optimally use all available data (53,64,65). Furthermore, unless the putative QTL is bracketed by two genetic markers, it is not possible using ANOVA to estimate linkage distance between the marker and the QTL, or to obtain an estimate of the QTL effect unbiased by recombination. This can be accomplished by maximum likelihood (ML) techniques (64,66). It has also been shown that ML has more power than ANOVA to detect QTL effects for situations of incomplete linkage (53,65). The disadvantages of ML are that its application is significantly more complicated than ANOVA and suitable algorithms for most situations are not available in general-use statistical packages. Within-sire marker effect can also be detected by a chi-squared test (55). Of the three methods described above, chi-squared is the only one for which statistical power can be readily estimated (52).

For populations with good recording systems, power can be further increased by the technique of selective genotyping (53,67). If individuals with extreme values for quantitative traits are selected for genotyping, power to detect a segregating QTL is significantly increased, as opposed to random selection. Selective genotyping has been investigated for crosses between inbred lines, but not for either the daughter or granddaughter designs.

Statistical methods have been developed to map QTL with the aid of genetic markers based on linear model analysis and maximum likelihood (53,64,65). Linear model analysis can be used to map QTL only if putative QTL is bracketed by two genetic markers. Even in the latter case, several hundred individuals must be scored to obtain accurate estimates for the effect and location of a QTL of a magnitude less than one phenotypic standard deviation.

Within-Breed Genetic Manipulation of Quantitative Trait Loci

Marker-assisted selection within a breed can contribute to genetic progress in three ways:

1. By increasing selection intensity
2. By decreasing generation interval
3. By increasing accuracy of evaluation

An example from dairy cattle selection will be used to illustrate how within-breed MAS can increase selection intensity and decrease generation interval.

In most advanced dairy breeding programs, sires for general service are selected by progeny test (14,18). Genetic advancement per year is equal to the sum of genetic advancement per generation by the four paths of inheritance (sire to sire, sir to dam, dam to sire, and dam to dam) divided by the sum of the generation intervals. At the age of 1 year, young sires are mated to a limited number of test cows. By the time the daughter records on performance traits have been collected and analyzed, the sires are at least 5 years old. Currently, selection of sires for progeny test is based only on pedigree data. In many countries, selection intensity of young sires is low, due to the expense involved in progeny testing and keeping sires until daughter records become available.

Kashi et al. (68,69) suggests the following scheme to increase both accuracy of evaluation and selection intensity. Once segregating QTL for economic traits have been identified in the population, young sires would be screened for genetic markers, and those found with the most favorable genotypes would then be progeny tested. They assumed that the number of sires progeny-tested would not change; thus both the accuracy of the evaluations of the young sires and the selection intensity would be increased. The selection intensity in the dam-to-sire path would be reduced, because more cows would be used as potential bull dams. However, this reduction is likely to be minimal, because selection intensity is already very high in the dam-to-sire path and is increasing due to the practice of multiple ovulation and embryo transplant techniques. They estimated the increase in the rate of genetic progress as 25–50% of the current program.

Another alternative would be to eliminate progeny testing of young sires completely and select candidate sires only on their genotype for the segregating genetic markers. This would significantly decrease the generation interval in the sire-to-dam path, but may reduce the accuracy of sire evaluations. Sires of sires would still be selected, based on daughter performance. Several countries, Norway for example, currently do not select sires for general use based on progeny tests, but instead inseminate all cows with a group of young sires selected only on the evaluations of their sires and dams. Owen (70) showed that this procedure is about as efficient as the progeny test system. Thus a system that would incorporate the positive aspects of both schemes should be more efficient than either alternative. We are not aware if any estimates of the expected gain from this approach have been computed.

Smith and Simpson (71) used selection index theory to combine individual and sib performance information with marker information. They used this approach to estimate the expected genetic advancement from MAS due to more accurate genetic evaluation. They found that gains were minimal for high heritability traits, but significant for low heritability traits. It should be noted, however, that the assumptions underlying their analysis did not correspond to the typical situation with field records, where related animals are distributed nonorthogonally over nongenetic factors such as herds and seasons.

In the dairy cattle example used here, the increase in selection intensity and the decrease in generation interval were due to genetic evaluation of animals using marker information. Marker information is potentially available at birth of the animals, whereas progeny information becomes available much later and often after much expense. Thus, with marker information, a larger number of animals can be evaluated for selection at an earlier age. This results in an increase in selection intensity and a decrease in generation interval. If the marker information is combined with phenotypic information on relatives, more accurate evaluations could be obtained than with either type of information used alone. Thus, to maximize the benefit from MAS, marker information should be combined with phenotypic information in an optimum way.

Fernando and Grossman (72) have shown how to evaluate animals by BLUP with marker and phenotypic information. This procedure is based on the following mixed linear model:

$$y_i = x'_i\beta + v_i^p + v_i^m + u_i + e_i \qquad (5)$$

where y_i is the quantitative trait value of the ith animal, β is a vector of fixed effects of nongenetic factors such as herds and seasons, x'_i is a known vector relating β to y_i, v_i^p and v_i^m are the additive effects of the QTL alleles linked to the paternal and maternal markers, u_i is the residual additive effect of the remaining QTL alleles, and e_i is a random error. Let y be the vector of y_i's, v be defined as

$$v' = \left[v_1^p, v_1^m, v_2^p, v_2^m, \ldots, v_n^p, v_n^m\right]$$

u the vector of u_i's, and e the vector of e_i's.

The vectors v, u, and e are assumed to have null expectations, and thus $E(y) = X\beta$. Also denoted the $Var(v) = G_v$, $Var(u) = G_v$, and $Var(e) = I\sigma_e^2$, where I is an identity matrix of order n, the number of animals. The random vectors v, u, and e are assumed to be independent. The assumption that v and u are independent is equivalent to the assumption that the marker locus is not linked to the other QTL. Fernando and Grossman (72) have described a tabular method to construct G_v, which is a function of the marker information, the recombination rate (r), and $\sigma_v^2 = Var(v_i^p) = Var(v_i^m)$. The covariance matrix of the residual additive effects can be constructed using the well-known tabular method described by many authors, for example, Henderson (16). This matrix is a function of the relationship information and $\sigma_u^2 = Var(u_i)$.

Now, Eq. (5) can be written in matrix notation as

$$y = X\beta + Zv + u + e \tag{6}$$

where $Z = I \otimes 1'_2$, $1'_2 = [1, 1]$ and $\otimes$ stands for the kroneker product. Then, BLUP of v and u can be obtained as the solutions to

$$\begin{bmatrix} X'X & X'Z & X' \\ Z'X & Z'Z + G_v^{-1}\sigma_e^2 & Z' \\ X & Z & I + G_u^{-1}\sigma_e^2 \end{bmatrix} \begin{bmatrix} \tilde{\beta} \\ \tilde{v} \\ \tilde{u} \end{bmatrix} = \begin{bmatrix} X'y \\ Z'y \\ y \end{bmatrix} \tag{7}$$

Blup of a, the vector of additive genotypic values is

$$\tilde{a} = Z\tilde{v} + \tilde{u} \tag{8}$$

An algorithm for inverting G_v in linear time has been developed by Fernando and Grossman (72); G_u is inverted using Henderson's algorithm (16). Equation (7) can be solved by iterative techniques used to solve the usual MME (15,17).

As stated earlier, BLUP can be justified as an optimum procedure when the joint distribution of y and a is normal. Further, if normality is a good approximation, restricted maximum-likelihood estimates of the variance components ($\sigma_v^2, \sigma_u^2, \sigma_e^2$) and the recombination rate (r) can be obtained by maximizing the restricted likelihood function:

$$l(\sigma_v^2, \sigma_u^2, \sigma_e^2, r) = \exp[-1/2(y'y - \hat{\theta}'C\hat{\theta}] \\ |C|^{-1/2}|G_v|^{-1/2}(\sigma_u^2)^{-n/2}(\sigma_e^2)^{-n/2} \quad (9)$$

with respect to $\sigma_v^2, \sigma_u^2, \sigma_e^2$, and r, where $\hat{\theta}$ is a solution to Eq. (7), and |C| is the determinant of the coefficient matrix of Eq. (7). The determinant of C can be obtained by sparse matrix techniques (73). The determinant of G_v can be calculated using the identity

$$G_v^{-1} = QG_\epsilon^{-1}Q' \quad (10)$$

where Q is an upper triangular matrix with ones on the diagonal, and G_ϵ is a diagonal matrix, with elements being functions of σ_v^2 and r (72). The determinant of G_v^{-1} is

$$|G_v^{-1}| = |Q||G_\epsilon^{-1}||Q'| \\ = |G_\epsilon^{-1}|$$

because the |Q|=1. Thus, the determinant of G_v is

$$|G_v| = |G_\epsilon| = \prod_1^{2n} d_{ii}$$

where d_{ii} is the *ith* diagonal element of G_ϵ.

If only two alleles are segregating at the QTL locus that is linked to the marker locus, normality will be a poor approximation. The consequences of evaluation by BLUP when normality does not hold need to be examined.

Between-Breed Manipulation of Quantitative Trait Loci

We discussed above how in trait-based selection of progeny of a cross between two lines, desirable alleles would be lost by chance. If the specific QTL accounting for most of the variance between two strains were known a priori, strategies could be employed for the most efficient incorporation of desirable alleles into a group of selected individuals. Weller and Soller (74) studied strategies for incorporating a number of favorable genes into a single strain. They found the optimum strategy to be random mating among the different strains and mass selection on the number of positive alleles in each progeny. By this method, the most rapid gain is made in the first few generations. However, as shown by Smith and Simpson (71), selection on marker genotype is only slightly more effective than trait-based selection. It should be noted, though, that with trait-based selection, some of the favorable alleles will be lost after several generations. Using MAS it is possible to produce individuals homozygous for the favorable allele for a relatively large number of loci in 8–10 generations.

Although dominance, heterotic, and epistatic effects have been found for individual QTL (30,43,45), very little has been published about the utilization of this source of variation in breeding programs. Genotyping of QTL-linked genetic markers may allow for utilization of nonadditive sources of genetic variance, especially in between-breed crosses.

Selection of domestic animals is nearly always based on an index of several traits. Falconer (75) noted that during the course of a breeding program, negative genetic correlations will develop among the traits under selection. Those alleles with positive effects for both traits will be the first to undergo fixation, while those alleles with positive effects on some traits and negative effects on others will remain at intermediate frequency. Genetic progress for an index consisting of traits with major negative genetic correlations will be slow even if all traits have high heritability. Since, a priori, there is no reason to assume that the negative genetic correlation will be distributed evenly over all QTL affecting the traits in question, it should be possible to locate specific loci with positive effects on both traits. Selection for these loci may be significantly more effective than traditional selection index. Previous studies have found QTL with effects on pairs of traits in opposite direction to the trait-based genetic correlation (30). At present no study has estimated the gain expected by MAS in the case of index selection.

Introgression is less common in domestic animals than in plant species, because of limited fertility, longer generation intervals, and the greater expense of each individual. Nevertheless, introgression has been applied in domestic animals for genes with major economic impact. Notable examples are the Booroola gene for increased fecundity and a recently started program to introgress genes for trypanotolerance from N'Dama cattle into other tropical breeds (24,76).

Without utilizing genetic markers, introgression can be performed only if the genotype of the locus in question can be readily ascertained in the progeny of the different crosses. However, if the locus being introgressed is linked to a genetic marker, then even alleles without a readily identifiable phenotype can be transferred from one strain to another. This process is dramatically facilitated if the QTL is bracketed on either side by segregating genetic markers. If the recombination frequency between the two genetic markers is low, then it can be assumed that double recombinants will be extremely rare, and by selecting on the desired parental genotype, the QTL effect will be transferred nearly intact. Even for genes of major effect, such as the Booroola gene, linkage to segregating genetic markers would allow for earlier identification than possible by genotyping based on quantitative trait expression.

If the cross is made between two lines with a saturated RFLP map, then the process can be further accelerated by selecting those progeny with both the desired allele for the locus being transferred and the greatest number of alleles from the recurrent strain for the other loci (32,35,38). If individuals are selected at random from the background genome, about 94% of the genome ($1-0.5^4$) will will be from the recurrent parent after four generations of backcross. It is likely that by selection on genetic markers, nearly the same level could be reached after only three generations, with reasonable levels of selection intensity.

SUMMARY

Traditional trait-based breeding methodologies utilize only a fraction of the genetic variation present in outbred populations. Statistical techniques based on ANOVA and maximum likelihood have been developed to detect and map quantitative trait loci by linkage to genetic markers in crosses between inbred lines. Quantitative trait loci in outbred populations can be detected by analysis of large half-sib families. Selective genotyping and marker bracketing of quantitative trait loci increase the power of detection. Currently used methods to estimate breeding values from field data have been modified to incorporate QTL-linked marker genotype data. Selection schemes based on individual quantitative trait loci may be able to increase the accuracy of genetic evaluations, increase selection intensity, decrease generation intervals, and utilize epistatic and dominance genetic variation. Genetic markers will also aid in introgression of desirable genes from exotic breeds and breeding for selection indices composed of traits with negative genetic correlations.

REFERENCES

1. Goffinet B. Selection on selected records. Genet Sel Evol 1983; 15:91-98.
2. Fernando RL. Selection and assortative mating. Ph.D. thesis, University of Illinois, 1984.
3. Fernando RL, Gianola D. Optimal properties of the conditional mean as a selection criterion. Theor Appl Genet 1986; 72:822-825.
4. Bulmer MG. The mathematical theory of quantitative genetics. Oxford: Clarendon Press, 1980.
5. Henderson CR. Selection index and expected genetic advance. In: Hanson WD, Robinson HF, eds. Statistical genetics and plant breeding. Washington DC: Publication 982, National Academy of Sciences, National Research Council, 1963:141-163.
6. Henderson CR. Sire evaluation and genetics trends. In: Anim. breed. genet. symp. in honor of Dr JL Lush. Champaign, Il: Amer Soc Anim Sci and Am Dairy Sci Assoc, 1973:10-41.
7. Henderson CR. Applications of linear models in animal breeding. Guelph: University of Guelph, 1984.

8. Henderson CR. Estimation of variance and covariance components. Biometrics 1953; 9:226-252.
9. Rao CR. Estimation of variance and covariance components—MINQUE theory. J Multivariate Anal 1971; 1:257-275.
10. Patterson HD, Thompson R. Recovery of inter-block information when block sizes are unequal. Biometrika 1971; 58:545-554.
11. Harville DA. Maximum likelihood approaches to variance component estimation and to related problems. J Am Stat Assoc 1976; 72:320-340.
12. Harvill DA. Computational aspects of likelihood-based inference for variance components. In: Gianola D, Hammond K, eds. Advances in statistical methods for genetic improvement of livestock. New York: Springer-Verlag, 1990.
13. Gianola D, Foulley JL, Fernando RL. Prediction of breeding values when variances are not known. In 3rd world congress genet. appl. livest. prod, Lincoln, NE, 1986; 12:356-370.
14. Weller JI, Ron M, Bar-Anan R. Multilactation genetic evaluation of the Israeli dairy cattle population, In 3rd world congress genet. appl. livest. prod, Lincoln, NE, 1986; 9:202-207.
15. Wiggans GR, Misztal I, Van Vleck LD. Implementation of an animal model for genetic evaluation of dairy cattle in the United States. In: Schmidt GH, ed. Proceedings of the animal model workshop. Champaign, IL: Am Dairy Sci Assoc, 1988:54-69.
16. Henderson CR. A simple method for computing the inverse of a numerator relationship matrix used in prediction of breeding values. Biometrics 1976; 32:69-83.
17. Misztal I, Gianola D. Indirect solution of mixed model equations. J Dairy Sci 1987; 70:716-723.
18. Van Vleck LD. Evaluation of dairy cattle breeding programs: specialized milk production. In 3rd world congress genet. appl. livest. prod. Lincoln, NE, 1986; 9:141.
19. Koch RM, Gibb JB, Gosey JA. Evaluation of beef cattle industry breeding programs: breeders and breed associations In 3rd world congress genet. appl. livest. prod. Lincoln, NE, 1986; 9:398-409.
20. Eikje ED. Studies on sheep production records. VIII. Estimation of genetic change. Acta Agric Scand, 1975; 25:253-260.
21. Weller JI, Wiggans GR, Lawlor TJ. Genetic evaluation of dairy goat does for milk and fat as an extension of buck evaluation. J Dairy Sci 1987; 70:681-686.

22. Kennedy BW, Hudson GFS, Schaeffer LR. Evaluation of genetic change in performance tested pigs in canada. In 3rd world congress genet. appl. livest. prod. Lincoln, NE, 1986; 9:149-154.
23. Moav R. Specialized sire and dam lines. I. Economic evaluation of cross-breeds. Anim. Prod. 1966; 8:193-202.
24. Piper LR, Bindon BM. The Booroola Merino and the performance of medium non-Peppin crosses at Armidale. Melbourne: CSIRO, 1982:9.
25. Carriquiry AL, Fernando RL, Gianola D, Grossman M. Effect of major genes on response to selection: a computer simulation. J Anim Sci 1984; 59(suppl 1):82-83.
26. Hoeschele I. genetic evaluation with data presenting evidence of mixed major gene and polygenic inheritance. Theor Appl Genet 1989; 76:81-92.
27. Soller M, Genizi A, Brody T. On the power of experimental designs for the detection of linkage between marker loci and quantitative loci in crosses between inbred lines. Theor Appl Genet 1976; 47:35-39.
28. Beckmann JS, Soller M. Detection of linkage between marker loci and loci affecting quantitative traits in crosses between segregating populations. Theor Appl Genet 1988; 76:228-236.
29. Tanksely SD, Medina-Filho R, Rick DM. Use of naturally-occurring enzyme variation to detect and map genes controlling quantitative traits in an interspecific backcross of tomato. Heredity 1982; 49:11-25.
30. Weller JI, Soller M, Brody T. Linkage analysis of quantitative traits in an interspecific cross of tomato. (*Lycopersicon esculentum* x *Lycopersicon pimpinellifolium*) by means of genetic markers. Genetics 1988; 118:329-339.
31. Soller M. Genetic mapping of the bovine genome using DNA-level markers with particular attention to loci affecting quantitative traits of economic importance. J Dairy Sci 1990; 73:2628-2646.
32. Bernatzky R, Tanksely SD. Toward a saturated linkage map of tomato based on isozymes and random cDNA sequences. Genetics 1986; 112:887-898.
33. Beckmann JS, Soller M. Restriction fragment length polymorphisms in genetic improvement: methodologies, mapping and costs. Theor Appl Genet 1983; 67:35-43.
34. Beckmann JS, Soller M. Molecular markers in the genetic improvement of farm animals. Biotechnology 1987; 5:573.
35. Paterson AH, Lander ES, Hewitt JD, Peterson S, Lincoln SE, Tanksley SD. Resolution of quantitative traits into mendelian factors by using a

complete linkage map of restriction fragment length polymorphisms. Nature 1988; 335:721-726.

36. Soller M, Beckmann JS. Restriction fragment length polymorphisms and genetic improvement. In Proc. 2nd world cong. appl. livest. prod., Madrid, 1982; 6:396.
37. Soller M, Beckmann JS. Restriction fragment length polymorphisms in poultry breeding: Poultry Sci. 1986; 65:1474-1488.
38. Tanksley SD, Young ND, Paterson AH, Bonierbale MW. RFLP mapping in plant breeding: new tools for an old science. Bio/Technology 1989; 7:257.
39. Beckmann JS, Kashi Y, Hallerman EM, Nave A, Soller M. Restriction fragment length polymorphisms among Israeli Holstein-Friesian dairy bulls. Anim Genet 1986; 17:25-38.
40. Fries R, Beckmann JS, Georges M, Soller M, Womack J. The bovine gene map. Anim Genet 1989; 20:3-29.
41. Hallerman EM, Nave A, Kashi Y, Beckmann JS. Restriction fragment length polymorphisms in dairy and beef cattle at the growth hormone and prolactin loci. Anim Genet 1987; 18:213-222.
42. Hallerman EM, Nave A, Soller M, Beckmann JS. Screening of Israeli Holstein-Friesian cattle for RFLP's using homologous and heterologous probes. J Dairy Sci 1988; 71:3378-389.
43. Edwards MD, Stuber CW, Wendel JF. Molecular-marker-facilitated investigations of quantitative trait loci in maize. I. Numbers, genomic distribution and types of gene action. Genetics 1987; 116:113-125.
44. Kahler AL, Wherhahn CF. Associations between quantitative traits and enzyme loci in the F2 population of a maize hybrid. Theor Appl Genet 1986; 72:15-26.
45. Stuber CW, Moll RH, Goodman MM, Shaffer HE, Weir BS. Allozyme frequency changes associated with selection for increased grain yield in maize. (*Zea mays* L.). Genetics 1988; 95:225-236.
46. Thoday JM. Location of polygenes. Nature 1961; 191:368-370.
47. Zhuchenko AA, Korol AB, Andryushchenko VK. Linkage between loci of quantitative characters and marker loci. Genetika 1979; 14:771-778.
48. Zhuchenko AA, Samovol AP, Korol AB, Andryushchenko VK. Linkage between loci of quantitative characters and marker loci. II. Influence of three tomato chromosomes on variability of five quantitative characters in backcross progenies. Genetika 1979; 15:672-683.
49. Arave CW, Lamb RC, Hines HC. Blood and milk protein polymorphisms in relation to feed efficiency and production traits of dairy cattle. J Dairy Sci 1971; 54:106-112.

50. Brum EW, Rausch WH, Hines HC, Ludwick TM. Association between milk and blood polymorphism types and lactation traits of Holstein cattle. J Dairy Sci 1968; 51:1031-1038.
51. Hines HC, Kiddy CA, Brum EW, Arave CW. Linkage among cattle blood and milk polymorphisms. Genetics 1969; 62:401-412.
52. Weller JI, Kashi Y, Soller M. Estimation of sample size necessary for genetic mapping of quantitative traits in dairy cattle using genetic markers. J Dairy Sci 1988; 71(suppl 1):142.
53. Lander ES, Botstein D. Mapping Mendelian factors underlying quantitative traits using RFLP linkage maps. Genetics 1989; 121:185-199.
54. Beever JS, George PD, Fernando RL, Stormont CJ, Lewin HA. Association between genetic markers and growth and carcass traits in a paternal half-sib family of Angus cattle. J Anim Sci 1990 68:337-344.
55. Gelderman H. Investigations on inheritance of quantitative characters in animals by gene markers. I. Methods. Theor Appl Genet 1975; 46:319-330.
56. Gelderman H, Peiper U, Roth B. Effects of marker chromosome sections on milk performance in cattle. Theor Appl Genet 1985; 70:138-146.
57. Goyon DS, Mather RE, Hines HC, Haenlein GFW, Arave CW, Gaunt SN. Associations of bovine blood and milk polymorphisms with lactation traits: Holsteins. J Dairy Sci 1987; 70:2585-2598.
58. Haenlein GFW, Goyon DS, Mather RE, Hines HC. Association of bovine blood and milk polymorphisms with lactation traits: Guernseys. J Dairy Sci 1987; 70:2599-2609.
59. Neimann-Sorenssen A, Robertson A. The association between blood groups and several production characters in three Danish cattle breeds. Acta Agric Scand 1961; 11:163-196.
60. Soller M. The use of loci associated with quantitative traits in dairy cattle improvement. Anim Prod 1978; 27:133-139.
61. Soller M, Genizi A. The efficiency of experimental designs for the detection of linkage between a marker locus and a locus affecting a quantitative trait in segregating populations. Biometrics 1978; 34:47-55.
62. SAS Institute. SAS user's guide: statistics, version 5. Cary, NC: SAS Institute, 1985.
63. Weller JI. Mapping and analysis of quantitative trait loci in Lycopersicon. Heredity 1987; 59:413-421.
64. Weller JI. Maximum likelihood techniques for the mapping and analysis of quantitative trait loci with the aid of genetic markers. Biometrics 1986; 42:627-640.

65. Simpson SP. Detection of linkage between quantitative trait loci and restriction fragment length polymorphisms using inbred lines. Theor Appl Genet 1989; 77:815–819.
66. Dempster AP, Laird NM, Rubin DB. Maximum likelihood from incomplete data via the EM algorithm. J Roy Stat Soc (series B) 1977; 39:1–38.
67. Lebowitz RJ, Soller M, Beckmann JS. Trait-based analyses for the detection of linkage between marker loci and quantitative trait loci in crosses between inbred lines. Theor Appl Genet 1987; 73:556–562.
68. Kashi Y, Soller M, Hallerman EM, Beckmann JS. Restriction fragment length polymorphisms in dairy cattle genetic improvement. In: 3rd world congress genet. appl. livest. prod. Lincoln, NE, 1986; 12:57–63
69. Kashi Y, Hallerman EM, Soller M. Marker-assisted selection of candidate sires for progeny testing programs. Anim Prod 1990; 51:63–74.
70. Owen JB. Selection of dairy bulls on half-sister records. Anim Prod 1975; 20:1–10.
71. Smith C, Simpson SP. The use of genetic polymorphisms in livestock improvement. J Anim Breed Genet 1986; 103:205.
72. Fernando R, Grossman M. Marker assisted selection using best linear unbiased prediction. Genet Sel Evol 1989; 21:467–477.
73. Misztal I. Restricted maximum likelihood estimation of variance components in animal model using sparse matrix inversion and a supercomputer. J Dairy Sci 1990; 73:163–172.
74. Weller JI, Soller M. Methods for production of multimarker strains. Theor Appl Genet 1981; 59:73–77.
75. Falconer DS. Introduction to quantitative genetics. 2nd ed. New York: Longman, 1981.
76. Soller M, Beckmann JS. Towards an understanding of the genetic basis of trypanotolerance in the N'Dama cattle of West Africa. Consultation report submitted to FAO, Rome, March 1987.

Index

About the Editors

LAWRENCE B. SCHOOK is Professor of Molecular Immunology in the Department of Animal Sciences at the University of Illinois at Urbana—Champaign. The author or coauthor of over 140 articles, book chapters, and abstracts, he is the editor of two books, including *Monoclonal Antibody Production Techniques and Applications* (Marcel Dekker, Inc.) and Editor-in-Chief of the *Animal Biotechnology* journal (Marcel Dekker, Inc.). Dr. Schook is a Charter Member of the Association of Medical Laboratory Immunologists and a member of the American Association for the Advancement of Science, American Society of Animal Science, American Association of Immunologists, American Society for Microbiology, American Society of Human Genetics, International Society for Animal Genetics, and Sigma Xi, among others. He received the B. A. (1972) degree from Albion College and M.S. (1975) and Ph.D. (1978) degrees from Wayne State University, Detroit, both in Michigan.

HARRIS A. LEWIN is Associate Professor of Immunogenetics in the Department of Animal Sciences at the University of Illinois at Urbana—Champaign. The author or coauthor of over 30 articles and book chapters, he is an Associate Editor of the *Animal Biotechnology* journal (Marcel Dekker, Inc.). A member of the American Association of Immunologists, American Association of Veterinary Immunologists, and International Society for Animal Genetics, Dr. Lewin received the B.S. (1979) and M.S. (1981) degrees from Cornell University, Ithaca, New York, and Ph.D. (1984) from the University of California, Davis.

DAVID G. McLAREN is Assistant Professor of Swine Breeding and Management in the Department of Animal Sciences at the University of Illinois at Urbana—Champaign. The author or coauthor of over 60 articles and abstracts, he is a member of the American Association for the Advancement of Science, American Society of Animal Science, and American Association of University Professors. Dr. McLaren received the Higher National Diploma (1980) degree from the Welsh Agricultural College, Aberystwyth, Wales, and M.S. (1982) and Ph.D. (1985) degrees from Oklahoma State University, Stillwater.

Milton Keynes UK
Ingram Content Group UK Ltd.
UKHW021630071024
449327UK00020BA/1261